The dynamics of sorptions processes

Динамика сорбционных процессов

Grundlagen der Verfahrenstechnik und chemischen Technologie

Основы химической техники и технологии

Herausgeber

Prof. Dr. Dr. E. h. K. Hartmann, Merseburg

Prof. Dr. W. Schirmer, Berlin
Ordentliches Mitglied der Akademie der Wissenschaften der DDR

Prof. Dr. M. G. Slinko, Moskau
Korr. Mitglied der Akademie der Wissenschaften der UdSSR

Редакторы

профессор доктор К. Хартманн, Мерзебург

профессор доктор В. Ширмер, Берлин
член АН ГДР

профессор доктор М. Г. Слинько, Москва
член-корреспондент АН СССР

AKADEMIE-VERLAG · BERLIN
1982

The dynamics of sorptions processes

Динамика сорбционных процессов

With 79 Figures and 5 Tables

AKADEMIE-VERLAG · BERLIN
1982

Verantwortlich für die Herausgabe dieses Bandes:
Prof. Dr. Dr. E. h. K. Hartmann, Merseburg

Erschienen im Akademie-Verlag, DDR - 1086 Berlin, Leipziger Straße 3—4
Lektor: Fritz Schulz

Lizenznummer: 202 · 100/462/82
Gesamtherstellung: VEB Druckhaus „Maxim Gorki“, 7400 Altenburg
Bestellnummer: 762 958 6 (2144/12) · LSV 1205/1095
Printed in GDR
DDR 42,— M

Contents

Содержание

Non-linear sorption kinetics equations

(I. N. TAGANOV)

Zusammenfassung:

Es wird eine verallgemeinerte Stofftransportgleichung der Form

$$\bar{\jmath}_m = -\int_0^\infty a(s)\, \bar{g}^{-n}(\tau - S)\, dS$$

$$n \leqq 1$$

$$\bar{g}(\tau - S) \equiv \nabla C(\tau - S)$$

eingeführt. Unter Verwendung dieser Gleichung werden verschiedene Varianten linearer und nichtlinearer Gleichungen der Sorptionskinetik abgeleitet.

The process of sorbent structure filling with molecules and ions in the sorption is usually described by employing the equations of diffusion and heat conductivity (in case of non-isothermic adsorption). In so doing non-isothermic adsorption is estimated with the help of the following equations [1, 2]

$$\left.\begin{aligned} \frac{\partial C}{\partial \tau} &= \nabla[D\nabla C + \delta_t \nabla t] - \frac{\partial a}{\partial \tau} \\ c_v \frac{\partial t}{\partial \tau} &= \nabla(\lambda \nabla t) + q_a \frac{\partial a}{\partial \tau} \\ a &= f(C, t) \end{aligned}\right\} \tag{1}$$

where:

C — sorbing substance concentration in the moving phase;
t — temperature;
a — the adsorbed substance concentration;
D — diffusion coefficient;
δ_t — thermodiffusion coefficient;
λ — coefficient of heat conductivity;
c_v — sorbent specific thermal capacity;
q_a — adsorption thermal effect;
$a = f(C, t)$ — adsorption thermal equation.

The initial conditions of equations (1) are of the following type;

$$\begin{aligned} C(0, \bar{r}) &= 0 \\ t(0, \bar{r}) &= t_0. \end{aligned} \tag{2}$$

Boundary conditions are determined by the relationship of intensity of internal and external diffusion. In convective heat — and mass-exchange at the border surface boundary conditions of (1) are the following

$$
\begin{aligned}
-D \left.\frac{\partial C}{\partial \bar{r}}\right|_s &= \beta_c(C_0 - C|_s) \\
-\lambda \left.\frac{\partial t}{\partial \bar{r}}\right|_s &= \beta_t(t_0 - t|_s)
\end{aligned} \tag{3}
$$

where:

β_t and β_c — heat- and mass-exchange coefficients respectively;
C_0, t_0 — concentration and temperature of the medium surrounding sorbent grains.

If the external diffusion rate is great in comparison with the internal diffusion one, the characteristic time of sorption and desorption processes on the surface being insignificant, the boundary conditions of (1) can be obtained by employing sorption equilibrium condition at the border surface:

$$
\begin{aligned}
a|_s &= f(C_0, t_0) \\
-\lambda \left.\frac{\partial t}{\partial \bar{r}}\right|_s &= \beta_t(t_0 - t|_s).
\end{aligned} \tag{4}
$$

In cases when temperature gradients in the sorbent grain are low and the diffusion coefficient does not depend on coordinates the equation is significantly simplified

$$
\begin{aligned}
\frac{\partial C}{\partial \tau} &= D \Delta C - \frac{\partial a}{\partial \tau} \\
a &= f(C).
\end{aligned} \tag{5}
$$

Applying the adsorption isotherm equation it is possible to exclude $\frac{\partial a}{\partial \tau}$ from (5)

$$
\begin{aligned}
\frac{\partial C}{\partial \tau} &= D(C) \Delta C \\
D(C) &= \frac{D}{1 + f'(C)}.
\end{aligned} \tag{6}
$$

Employing the inverted isotherm $C = f_*(a)$ permits to convert the equation to the following form

$$
\begin{aligned}
\frac{\partial a}{\partial \tau} &= D(a)\,[f''(a)\,(\nabla a)^2 + f'(a)\,\Delta a] \\
D(a) &= \frac{D}{1 + f_*'(a)}.
\end{aligned} \tag{7}
$$

In (6) and (7) prime denotes differentiation by argument. Equation (6) is linear only for the linear isotherm $a = \gamma C$ and $D(C)$ being equal to $\frac{D}{1 + \gamma}$.

Applying equation (6) provides the necessary accuracy in the description of the adsorption process, which is characterized by linear and close to linear isotherms. In cases when complex interaction of diffusing sustance and sorbent structure takes place various defections from the diffusion transfer mechanism are observed. Adsorption isotherms are estimated experimentally by integral methods as a rule and do not provide information on dynamic characteristics of interaction between sorbing substance and sorbent structure. Detailed investigation of substance transfer mechanisms in the sorbent and their interaction with the sorbent structure would result in the following relationship for the adsorption isotherm under dynamic conditions:

$$a = f\left(C, \nabla C, \frac{\partial C}{\partial \tau}, \ldots\right). \tag{8}$$

However, modern adsorption theory and experimental standard do not provide sufficiently universal dependence of (8) type and in non-linear effect analysis it is reasonable to apply methods of the phenomenological theory of mass-transfer.

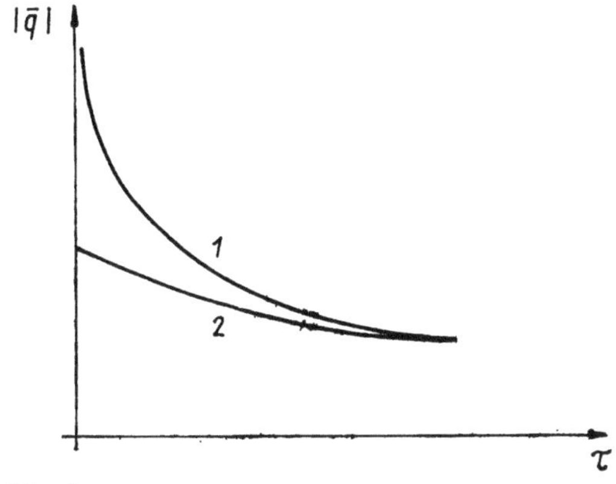

Fig. 1

Before going on to the derivation of non-linear transfer equations in the adsorbent it is appropriate to accomplish general non-linear effect classification.

As it is known the classical diffusion equation based on the transfer law in the form:

$$\bar{j}_c = -D\nabla C \tag{9}$$

assumes the possibility of mass distribution with indefinitely large velocity so that the solution of (6) with the boundary conditions of the first type results in continuous infinite flow through the surface corresponding to the initial conditions. (curve *1* in Fig. 1). This circumstance in its turn is likely to result in an abnormally steep rise sorption integral curve at initial periods of time. Actually, dependence of curve 2 type is observed in Fig. 1.

While analyzing the extent of diffusion flow dependence upon the concentration gradient in adsorption it may occur that this dependence does not follow linear law of (9) (curve *1* in Fig. 2) and possesses a monotonically decreasing curvature (curve *2* in Fig. 2) or even shows a tendency to saturation (curve *3* in Fig. 2). In a number of cases such effects can be explained by decreasing dependence of the diffusion coefficient upon the concentration. However, there are cases when the introduction of a diffusion varying coefficient fails to be a success. For example in solving reverse problems for sorbents of various geometrical shapes, when all other conditions are equal, various diffusion coefficient dependences upon concentrations are obtained.

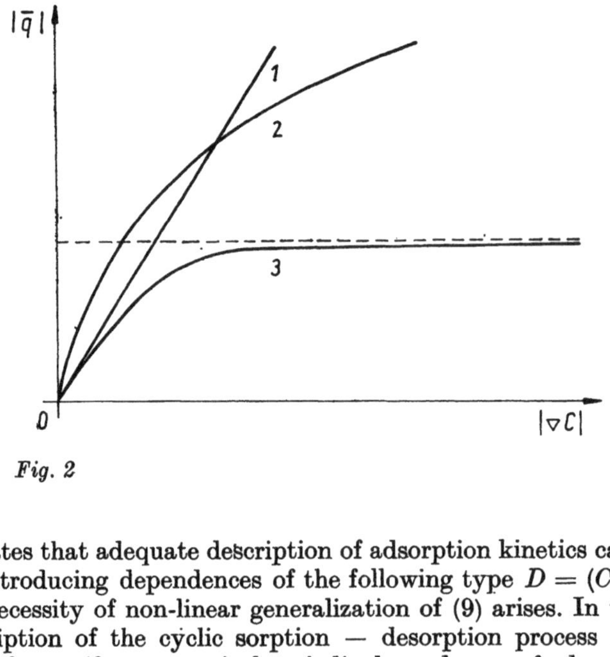

Fig. 2

This indicates that adequate description of adsorption kinetics can be attained only by introducing dependences of the following type $D = (C, \nabla C)$, i.e. the essential necessity of non-linear generalization of (9) arises. In the mathematical description of the cyclic sorption — desorption process by employing equations of type (6) symmetrical periodic dependences of adsorbed substance quantity C upon time are obtained (curve *1* in Fig. 3).

Actually, in many cases asymmetric curves with progressing modifications of periodically repeated area forms are experimentally observed (curve *2* in Fig. 3). This is connected with the fact that in repeating sorption and desorption phases a gradual modification of sorptive and diffusion properties of porous structure occurs. One of the possible approaches to describing such effects in sorption process is "genetic memory" introduction.

In recent years [3–5] in heat conductivity theory integral generalization of FOURIER law of the following type was investigated:

$$\bar{q}(\tau) = -\int_0^\infty a(s)\, \bar{g}(\tau - s)\, ds\,. \tag{10}$$

Here:

$\bar{q}(\tau)$ — heat flow;
$\bar{g}(\tau - s) = \nabla t(\tau - s)$ temperature gradient.

The generalization permits to consider effects similar to the ones mentioned above. In particular it was shown that equation (10) is adequate for the description of the heat distribution process in the materials with memory. If $a(s) = k\sigma e^{-\sigma s}$ (10) converts into MAXWELL-CATTANEO [6, 7] ratio:

$$\bar{q} = -k\nabla t - \frac{1}{\sigma}\frac{\partial \bar{q}}{\partial \tau} \tag{11}$$

providing finite velocity of perturbation of heat distribution.

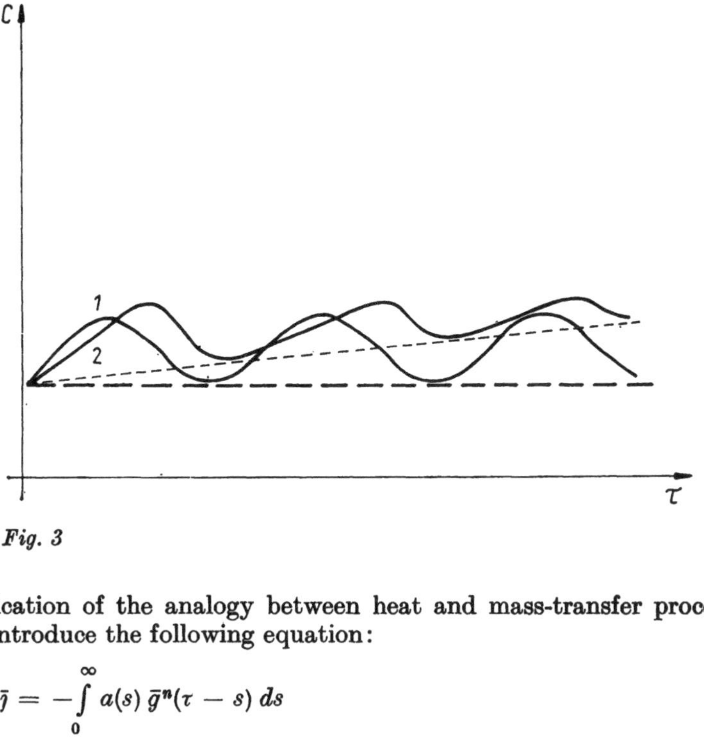

Fig. 3

The application of the analogy between heat and mass-transfer processes permits to introduce the following equation:

$$\bar{j} = -\int_0^\infty a(s)\, \bar{g}^n(\tau - s)\, ds$$

$$n \leqq 1 \tag{12}$$

$$\bar{g}(\tau - s) \equiv \nabla C(\tau - s),$$

which makes it possible to obtain equations describing the substance transfer process in the sorbent with finite distribution velocity, nonlinear transfer flow dependence upon concentration gradient and "genetic memory" effects.

If the concentration gradient does not depend upon s, the substance flow value is estimated by employing the instantaneous concentration gradient.

The following expression can be obtained from (12)

$$\bar{j} = -D^*(\nabla C)^n \qquad n \leq 1, \qquad D^* = \int_0^\infty a(s)\, ds \tag{13}$$

When $n < 1$ equation (13) describes the transfer with flow dependence upon the concentration gradient of curve type 2 in Fig. 2. The corresponding equation of substance transfer in the adsorbent is:

$$\frac{\partial C}{\partial \tau} = D^*\nabla(\nabla C)^n - \frac{\partial a}{\partial \tau} \qquad a = f(C);\ n < 1 \tag{14}$$

or after $\frac{\partial a}{\partial \tau}$ having been excluded:

$$\frac{\partial C}{\partial \tau} = \frac{D^*}{1 + f'(C)} \nabla(\nabla C)^n \qquad n < 1 \tag{15}$$

In the absence of concentration gradient dependence upon s in (12) and expression for $a(s)$:

$$a(s) = \frac{D^*}{\tau^*} e^{-\frac{s}{\tau^*}} \tag{16}$$

equation (12) becomes:

$$\dot{j} = -D^*(\nabla C)^n - \tau^* \frac{\partial \bar{j}}{\partial \tau}. \tag{17}$$

Such an equation describes the substance transfer with the sharply limited front moving with finite velocity and non-linear dependence of flow upon the concentration gradient. The corresponding equation of substance transfer in the sorbent is:

$$\frac{\partial C}{\partial \tau} + \tau^* \frac{\partial^2 C}{\partial \tau^2} = D^*\nabla(\nabla C)^n - \frac{\partial a}{\partial \tau} \qquad a = f(C);\ n < 1 \tag{18}$$

In case of the linear isotherm $a = \gamma C$ equation (18) is simplified.

$$\frac{\partial C}{\partial \tau} + \frac{\tau^*}{1+\gamma} \frac{\partial^2 C}{\partial \tau^2} = \frac{D^*}{1+\gamma} \nabla(\nabla C)^n \qquad n < 1 \tag{19}$$

When $n = 1$ equation (18) converts into the sorption kinetics hyperbolic equation:

$$\frac{\partial C}{\partial \tau} + \tau^* \frac{\partial^2 C}{\partial \tau^2} = D^*\Delta C - \frac{\partial a}{\partial \tau} \qquad a = f(C) \tag{20}$$

In solving concrete problems by employing equations (18)—(20) it is necessary to add to the initial conditions of type (2) the definition of the value $\frac{\partial C}{\partial t}$ at the initial moment of time.

Direct application of equation (12) leads to non-linear integro-differential sorption kinetics equation:

$$\tau^* \frac{\partial^2 C}{\partial \tau^2} + b(0) \frac{\partial C}{\partial \tau} + \int_0^\infty b'(s) \frac{\partial}{\partial \tau} C(\tau - s)\, ds$$
$$= D^* \nabla (\nabla C)^n + \int_0^\infty a'(s)\, \nabla[\nabla C(\tau - s)]^n\, ds - \frac{\partial a}{\partial \tau}; \tag{21}$$
$$n \leqq 1;\ a = f(C),$$

where:

$b(s)$ — function of chemical potential relaxation;
$a(s)$ — function of concentration gradient relaxation.

Experimental methods and methods of reverse problem solving for the determination of relaxation functions $a(s)$ and $b(s)$ have not been developed and at present equation (21) seems to be applied only to quality analysis of the memory effects in sorption-desorption processes.

The boundary conditions of (3) describing convective mass-exchange at the boundary surface in applying to equations (14) and (18) are modified:

$$-D^* \left(\frac{\partial C}{\partial \bar{r}}\right)^n \Bigg|_s = \beta_c (C_0 - C|_s). \qquad n \leqq 1 \tag{22}$$

It is interesting to note that the equation of type (20) can be obtained from (5) if it is assumed that dependence (8) is:

$$a = \varphi(C) + \tau^* \frac{\partial C}{\partial \tau}. \tag{23}$$

Substituting (23) into (5) results in the following equation:

$$\frac{\partial C}{\partial \tau} + \frac{\tau^*}{1 + \varphi'(C)} \frac{\partial^2 C}{\partial \tau^2} = \frac{D}{1 + \varphi'(C)} \Delta C \tag{24}$$

Frequently it is reasonable to represent and analyse adsorption kinetics problem solving in dimensionless form. Equation (18) can be represented in dimensionless form after the convecting term having been added for generality. The original equation with initial and boundary conditions is one-dimensional:

$$\left.\begin{aligned} &\frac{\partial C}{\partial \tau} + v \frac{\partial C}{\partial x} + \tau^* \frac{\partial^2 C}{\partial \tau^2} = D^* \frac{\partial}{\partial x}\left[\left(\frac{\partial C}{\partial x}\right)\right]^n \\ &C|_{\tau=0} = F(x) \qquad \frac{\partial C}{\partial \tau}\Big|_{\tau=0} = \text{const} \\ &-D^* \left(\frac{\partial C}{\partial x}\right)^n \Bigg|_s = \beta (C|_s - C_0) \end{aligned}\right\} \tag{25}$$

After introducing spatial and time scales L and T it is possible to represent system (25) in dimensionless form:

$$\left.\begin{aligned} &\frac{1}{F_0^*}\frac{\partial c}{\partial \tau} + \frac{1}{\mathrm{Pe}^*}\frac{\partial c}{\partial x} + \frac{1}{\Omega^*}\frac{\partial^2 c}{\partial \tau^2} = \frac{\partial}{\partial x}\left[\left(\frac{\partial c}{\partial x}\right)^n\right] \\ &c|_{\tau=0} = F \qquad \left.\frac{\partial C}{\partial \tau}\right|_{\tau=0} = \mathrm{const} \\ &-\left.\left(\frac{dc}{\partial x}\right)^n\right|_s = Sh^*(c|_s - c_0) \end{aligned}\right\} \tag{26}$$

Here modified criteria are used: FOURIER number $F_0^* = \frac{D^*T}{L^{n+1}}$, PECLET number $\mathrm{Pe}^* = \frac{D^*}{vL^n}$, SHERWOOD number $Sh^* = \frac{\beta L^n}{D^*}$ criterion $\Omega^* = \frac{D^*\tau^*}{L^{n+1}}$. In order to solve adsorption kinetics non-linear equations various numerical methods of non-linear equation solution referred to in the special literature can be applied.

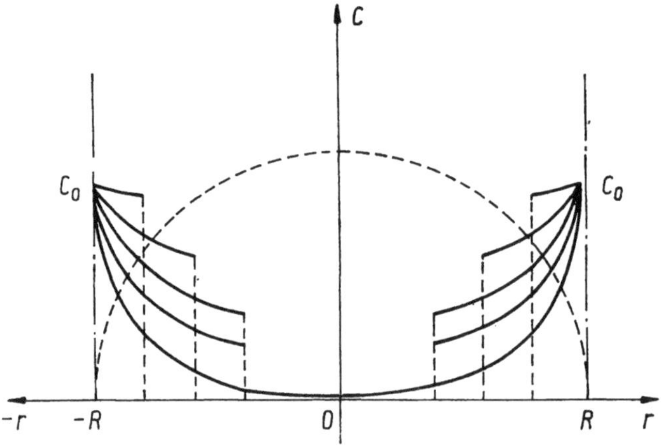

Fig. 4: Concentration distributions in the spherical particle of adsorbent at different moments.

It is noteworthy that the practice of concrete problem solution resulted in the conclusion that the solution of non-linear kinetics equation by employing digital computers is more laborious than that of linear ones because of worse convergence and stability of integration numerical methods.

Fig. 4 shows by way of example sorption kinetics non linear equation solution of type (18) for a spherical sorbent when $n = 1/2$ and with boundary condition (22). Concentration curves for increasing time moments τ_1, τ_2 and so on are characterized by sharply limited fronts moving towards the sorbent centre. Sorption kinetics hyperbolic equation (20) for linear isotherm can be solved analytically by the variables separation method.

For a spherical sorbent the following equation system can be considered:

$$\left.\begin{aligned}
&\frac{\partial C}{\partial \tau} + \frac{\tau^*}{1+\gamma}\frac{\partial^2 C}{\partial \tau^2} = \frac{D^*}{1+\gamma}\left(\frac{\partial^2 C}{\partial r^2} + \frac{2}{r}\frac{\partial C}{\partial r}\right)\\
&C(r, 0) = 0, \qquad \left.\frac{\partial C}{\partial \tau}\right|_{\tau=0} = 0\\
&C(R, \tau) = C_0, \qquad \left.\frac{\partial C}{\partial r}\right|_{r=0} = 0
\end{aligned}\right\} \tag{27}$$

The total quantity of adsorbed substance depending upon time is of practical interest rather than concentration fields:

$$C(\tau) = \frac{1}{V}\int_0^R C(r, \tau)\, \pi^2 r^2 dr. \tag{28}$$

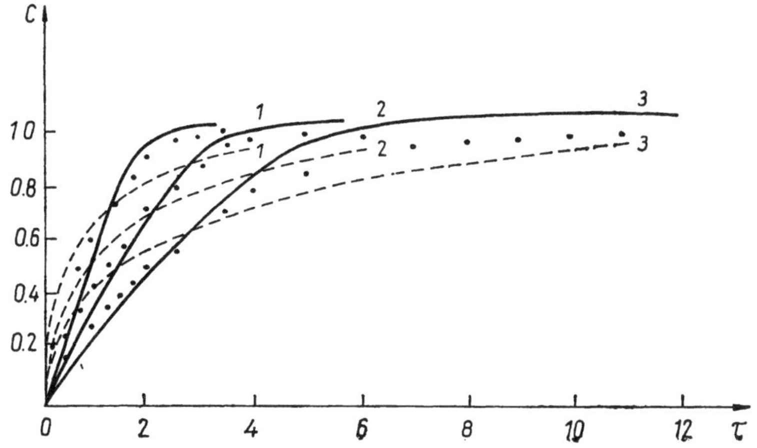

Fig. 5: Time dependence of the total quantity of adsorbed benzene for a spherical particle of coal (d = 2 mm) for different concentrations of benzene in the atmosphere: *1* — 70 mg/*l*, *2* — 34,6 mg/*l*, *3* — 18,3 mg/*l*. Dots-experimental data, solid curves for (27), dotted curves for (5).

Fig. 5 shows a comparison of kinetic curves calculated by employing the system (27) (solid curves) and kinetic curves calculated with the help of equation (5) (dotted curves) with experimental data. As it is seen from the figure the hyperbolic equation defines the sorption internal diffusion more precisely. The classical diffusion equation gives too steep a rise of kinetic curves for initial periods of time and on the whole poorly corresponds to experimental data.

References

[1] TUNITSKY, N. N., KAMINSKY, V. A., TIMASHEV, S. F., "*Methods of physico-chemical kinetics*", Chemistry, Moscow 1972.
[2] GOLUBEV, V. S., PANIENKOV, G. M., "*Theories of sorption dynamics and ionic exchange*", Coll. of works, "*Catalytic process kinetics*", Chemistry, Moscow 1969.
[3] GINSTIN, M. E., PIPKIN, A. C., Arch. Rat. Mech. Anal. **31**, 113–126 (1968).
[4] NUNZIATO, I. W., Quart. of Appl. Math., July, 187 (1971).
[5] LYKOV, A. V., IFG **26**, N5 (1974).
[6] MAXWELL, I. C., Phil. Trans. Roy. Soc., London **157**, 49–88. 1867.
[7] CATTANEO, C., Atti del Seminario matematico e fisico della Universita di Modena **3**, 3–21 (1948).

The formation of quasi-stationary fronts during the process of adsorption in a fixed bed

(V. L. Kolin)

Zusammenfassung:

Es wird die Bildung quasistationärer Konzentrationsfronten bei isothermer und nicht-isothermer Adsorption behandelt. Dazu werden numerische Methoden zur Lösung von Differentialgleichungssystemen hyperbolischen und parabolischen Typs diskutiert. Beispiellösungen werden angeführt und die qualitativen Besonderheiten der Konzentrationsfelder in Abhängigkeit von der Krümmung der Adsorptionsisotherme, den kinetischen Gesetzmäßigkeiten, dem Einfluß von Längsvermischungseffekten, Wärmeeffekten und Wärmeübertragungsbedingungen untersucht. Für das asymptotische Verhalten werden mathematische Modelle vorgeschlagen, die durch Systeme gewöhnlicher Differentialgleichungen repräsentiert werden. Es wird gezeigt, daß die Modellösungen befriedigend mit den experimentellen Daten sowohl für die Bildung der quasistationären Sorptionsfront als auch für das asymptotische Verhalten übereinstimmen.

1. Introduction

The theory of the dynamics of sorption concerns itself with the distribution in time of the concentrations through the layer of sorbent. This theory is the basis for engineering calculations, design engineering and devising optimum conditions for separation processes with components in gaseous, vapourgouseous or liquid phases.

The solution of the principal problem of dynamics is reduced to the integration of a system comprising kinetic equations and the equations of material and thermal balances for each of the phases participating in the mass-exchange. Some longitudinal transfer which occurs in the carrying agent and granulated sorbent flows shall be accounted for by certain effective coefficients. In many cases (adsorption in fluid phase; adsorption in gas-flows when concentrations of the sorbate are small) one may ignore the influence of the heat given off during the adsorption on the equilibrium, dynamic and kinetic characteristics of the process and this process may be regarded as isothermal.

However, even this simplified approach to the description of the process leaves some mathematical difficulties which enforce the imposition of various limitations and assumptions for the solution of important practical problems.

For the non-equilibrium sorption which is of practical interest the most common assumptions are:

1. The effects of longitudinal transfer are negligible, otherwise it is possible to account for these effects by the introduction of effective kinetic coefficients.
2. Reduction of kinetic factors to a certain effective constant of diffusion.
3. Assumption that the front of sorption shifted itself sufficiently along the bed and the frontage is much shorter than the length of the area saturated to its full capacity (exhausted) with sorbate.

The most common isotherm for technological processes is convex towards the concentration axis $\left(\frac{da^*}{dc^*} > 0\right)$. For this isotherm both experimental [1] and theoretical [2, 3] research show the existence of quasi-stationary conditions, characterized by parallel transfer of the concentration front. The condition is a limit of development for the process when $x \to \infty$, $t \to \infty$ and is at present being rather thoroughly explored for individual substances and for mixtures.

Modern technology makes high demads on the purity of the end products, liquid and gaseous industrial wastes, initial substances used for synthesis and other gaseous and liquid media with their compounds.

These demands stipulate the decrease in initial and permissible "slip through" concentrations in sorption processes.

In many vapour-gaseous and liquid-phase processes of separation or clearing the permissible concentration on leaving the bed is $(10^{-3} \div 10^{-5}) \cdot C_0$, or less. The required degree of clearing and sorbent capacity are usually attained by combining the specific interaction of the molecules which are being adsorbed with the solid surface and increase in the non-specific interaction due to the reduction of pore sizes. In some cases the ability for separation by sorption is determined by the molecular-sieve effect.

Decrease in pore sizes of the adsorbents till they are comparable to the effective diameter of the adsorbed molecules leads to the increase in influence of the intradiffusional coefficient which may alter depending on the sorption values and on the concentration of sorbate in the carrier flow.

In these cases the formal kinetic equation is usually employed though its use is only completely warranted for when the rate of the process is controlled by external diffusional resistance.

For most known solutions of the dynamics problem the kinetic coefficients are considered constant regardless of the stage of the process (external or internal mass-exchange or both) which determines the total diffusional resistance.

These circumstances compel one to take some objective factors into consideration, which seemed previously insignificant, when conducting analysis and calculations for a range of sorption processes. Thus a concept of the formed quasi-stationary front of sorption being much shorter than the length of the exhausted layer is rather frequently unsupported by experimental data. In many cases of fine purification of gaseous, vapour-gaseous and liquid streames the bed works with the sorption front yet unformed in the practically most interesting region of small "slip through" concentrations. Low rates of mass-exchange, peculiar to the sorption in liquid phase are the material factor for the expanding of the formation area. At the same time with the front of sorption unformed modern high-efficiency sorbents allow for such degree of utilization of the sorbent capacity which render it quite sufficient for industrial purposes. Therefore even with large-scale industrial processes it is sometimes unreasonable to acquire a bed of the length which provides for the forming of quasi-stationary fronts and is feasible for calculations based on equations by SHILOV [1], or MICHAELS-TREYBAL [4].

2. The beginning of the process

Let us submit for consideration an approximate mathematical model of the dynamics of isothermal adsorption in a layer of immobile granulated, grained or fibrous sorbing agent. The model consists of the following equations:

$$\varepsilon \frac{\partial c}{\partial t} = D_l \frac{\partial^2 c}{\partial x^2} - v \frac{\partial c}{\partial x} - (1 - \varepsilon) \frac{\partial a}{\partial t} \tag{1}$$

$$\frac{\partial a}{\partial t} = f(c, a) \tag{2}$$

of boundary and initial conditions

$$vc_0 = vc(0^+, t) - D_l \cdot \frac{\partial c(0^+, t)}{\partial x}; \quad \frac{\partial c(l^-, t)}{\partial x} = 0; \tag{3}$$

$$a(x, 0) = 0; \; c(x, 0) = 0; \; c(0^-, t) = c_0 = \text{const}. \tag{4}$$

The form of the function $f(c, a)$ may vary a great deal, depending on which stage is in charge of the process rate and on the equation for the adsorption isotherm.

For the solution of practical problems of the sorption dynamics one usually employs equations of kinetics that correspond to the highest degree to the experimental data. The following will be sufficient:

$$f(c, a) = \begin{cases} \beta_0[c - c^*(a)] & (5) \\ \beta_0[a^*(c) - a] & (6) \\ \beta_0 \cdot c \left(1 - \dfrac{a}{a_\infty}\right)^{\mu} & (7) \\ \beta_0 \left[c(1 - \theta) - \dfrac{1}{b}\,\theta\right] & (8) \end{cases}$$

Most methods for analysis and calculation of sorption processes are based on the assumption that the influence of longitudinal effects either is small or can be approximately accounted for by introduction of the effective coefficient of mass-exchange, as in the method offered by Todes and Bicson [5]. This makes it possible to substitute (approximately) for a system of parabolic type equations in partial derivatives (1), (2) a system of hyperbolic type equations

$$\varepsilon \frac{\partial c}{\partial t} = -v \frac{\partial c}{\partial x} - (1 - \varepsilon) \frac{\partial a}{\partial t}, \tag{9}$$

$$\frac{\partial a}{\partial t} = \beta_{eff} \cdot f_1(c, a). \tag{10}$$

At the stage of parallel transfer of the concentrations front the coefficient β_{eff}, according to [5] is found from the equation

$$\beta_{eff} = K_{D_2} \cdot \beta_0,$$

where

$$K_{D_2} = \frac{1}{1+D}; \quad \beta_0 = \frac{\beta_1 \cdot \beta_2}{\beta_1 + \beta_2}; \quad D = \frac{D_l \cdot \beta_0}{v^2}.$$

This approximate method of accounting for the influence of longitudinal effects is successfully employed for analysis and calculation for sorption processes which proceed under the condition of longitudinal transfer of the quasi-stationary front. However, when the process is a slow one or the layer of granular or fibrous adsorbent is thin, then there is not enough working-time for the condition of parallel transfer to establish itself and in these cases it is necessary to consider the initial stage of the sorption dynamics.

As it is shown by [6] for the initial stage of the process of sorption ("instantaneous" "slip through"), when $t = \frac{x}{v} \cdot \varepsilon$ the following equation is correct

$$\frac{C - C_H^*}{C_0 - C_H^*} = y = \exp\left[-\frac{2}{1+\sqrt{1+4D\cdot\alpha}} \cdot \frac{\beta_0 \cdot x}{v}(1-\varepsilon)\right]. \tag{11}$$

If we assume that the influence of the longitudinal transfer effects, resulting in the uncertain distribution of concentrations over the layer, vary monotonously with the development of the process from the initial stages to the stage of parallel transfer of the concentrations front, then, similar to [5] we may introduce for the initial stage

$$\beta_{eff} = \beta_0 \cdot \frac{2}{1+\sqrt{1+4D\alpha}} \tag{12}$$

which is $\beta_{eff} = K_{D_1} \cdot \beta_0$ where $K_{D_1} = \frac{2}{1+\sqrt{1+4D\alpha}}$. By this approach we can roughly describe the whole sorption process by the system (9), (10), if $\beta_{eff} = K_D\beta_0$ and $K_D \in [K_{D_2}; K_{D_1}]$. In employing for calculations of the dynamics of sorption the solutions for the system (9), (10) are applicable when $K_D = K_{D_1}$ and $K_D = K_{D_2}$. We create a certain "nook" where the real distributions of concentrations $c(x, t)$, $a(x, t)$ for all the stages of the sorption process will be contained. At the same time a contradiction arises out of the approximate substitution of the system (9), (10) for the system (1), (2). Indeed it follows from the system (1), (2) that at the asymptotic stage of the process, if the isotherm is convex, the distribution of the concentrations in the front of sorption should be such that $y < \gamma$.

At the same time, for the system (9), (10) approximately equivalent to (1), (2) at the asymptotic stage of the process that is $y \simeq \gamma$ is correct the correlation of Zeldovitch [3].

Analysis of the asymptotic stage of the process, described by the system (1), (2) leads to the following equation for the working line

$$\gamma = y + \frac{D_l \cdot \varepsilon^2}{\alpha v^2} \cdot \frac{(c_0 + \alpha a_0)^2}{c_0^2 \cdot a_0} \cdot \frac{dc}{dt}. \tag{13}$$

The second item of the expression (13) determines the influence of the longitudinal transfer effects on the balance of relative concentrations of the sorbate in various phases at a given point of the front of sorption and this item may be defined from the isoplanes of sorption. When $D_l \to 0$, (13) gives one the correlation by ZELDOVITCH. The higher D_l is the more the working line curves. All deviations from ZELDOVITCHS correlation depend on the rate of the process and on the isotherm of sorption. For sharply curved isotherms when the sorption rate is high, that is for steep exit curves and low v and c_0 values, this deviation grows. In order to evaluate the influence of D_l on the feasibility of ZELDOVITCHS correlation we shall examine the experimental data for two greatly differing cases: vapour adsorption out of an air stream and adsorption of a solute out of a stream of highly-viscous liquid. We calculate D_l from the correlation $D_l = 0.5v \cdot d$, which was introduced by LEVENSPIEL and BISHOFF [7]. For a case of moisture transferred in a liquid flow through a bed of small glass balls the results obtained by the author are: $D_l \simeq 1.6 . v \cdot d$.

The results obtained by calculations show that with typical sorption processes in vapour-gaseous and liquid phases the effect of longitudinal transfer on the deviations from ZELDOVITCHS correlation is not important.

Therefore one is justified in accounting for the influence of longitudinal effects with the effective kinetic coefficient

Influence of longitudinal transfer on the feasibility of ZELDOVITCHS correlation

y	Adsorption of benzene vapour out of air by active carbon	Adsorption of moisture out of synthetic oil by zeolite.
0.05	0.0502	0.05415
0.1	0.1003	0.1068
0.2	0.2048	0.2011
0.3	0.3058	0.3013
0.4	0.4075	0.4015
0.5	0.5088	0.5015
0.6	0.6099	0.6016
0.7	0.7015	0.7015
0.8	0.8100	0.8013
0.9	0.9085	—
0.95	0.9558	—

It is obvious, however, that longitudinal effects may influence β_{eff} to a considerable extent and thus influence the distribution in time of concentrations over the length of the bed. If we compare K_{D_1} and K_{D_2} we shall see that at the initial stage of the process the influence of D_l is less than at the stage of the parallel transfer of the quasi-stationary wave. The value of D grows with the growth of β_0 and decrease in ϑ, so the relative influence of parallel transfer is stronger for the processes characterized by higher values of kinetic coefficients.

3. Analytical solution of the problem of the dynamics of sorption

Through the simplifications introduced above for some kinetic equations we may obtain on analytical solution of the problem of dynamics.

So, mathematical model (9), (7) in a dimensionless form is

$$\frac{\partial y}{\partial \xi} + \frac{\partial \gamma}{\partial \tau} = 0 \ldots \tag{14}$$

$$\frac{\partial \gamma}{\partial \tau} = y(1 - \gamma)^{\mu} \tag{15}$$

Here

$$\tau = \frac{\beta_{eff}}{G}\left(t - \varepsilon \frac{x}{v}\right), \quad \xi = \frac{\beta_{eff}}{v}(1 - \varepsilon)\, x; \quad G = \frac{a_\infty}{c_0}$$

$$y = \frac{c}{c_0}; \quad \gamma = \frac{a}{a_\infty}$$

Dividing the variables in (15) and differentiating both parts of the equation by ξ we get the following result:

$$\frac{1}{\mu - 1} \cdot \frac{\partial^2}{\partial \xi\, \partial \tau}\left(\frac{1}{(1 - \gamma)^{\mu - 1}}\right) = \frac{\partial y}{\partial \xi} \tag{16}$$

Substituting in (16) $\frac{\partial y}{\partial \xi}$ for $-\frac{\partial \gamma}{\partial \tau}$ and integrating over τ we obtain:

$$\frac{1}{\gamma \cdot (1 - \gamma)^{\mu}} \cdot \frac{\partial \gamma}{\partial \xi} = -\gamma + \beta \tag{17}$$

where the integration constant $B(\xi) = 0$.

As $\frac{1}{\gamma(1 - \gamma)^{\mu}} = \sum_{k=0}^{\infty} (1 - \gamma)^{k-\mu}$, the result of the integration of (17) is

$$\sum_{k=0}^{\infty} \frac{(1 - \gamma)^{k-(\mu-1)}}{k - (\mu - 1)} = \xi + A(\tau) \tag{18}$$

The integration constant $A(\tau)$ may be obtained as a solution of the system (14), (15), when $\xi = 0$. For the frontal sorption process, when $y(0, \tau) = 1$ and $\gamma(\xi, 0) = 0$, it follows from (15) that

$$\gamma(0, \tau) = 1 - [1 + (\mu - 1)\, \tau]^{-\frac{1}{\mu - 1}} \tag{19}$$

Then, from (18) and taking (19) into consideration we get

$$A(\tau) = \sum_{k=0}^{\infty} \frac{[1 + (\mu - 1)\,\tau]^{1-\frac{k}{\mu-1}}}{k - (\mu - 1)} \tag{20}$$

Hence,

$$\xi = \sum_{k=0}^{\infty} \frac{(1-\gamma)^{k-(\mu-1)} - [1 + (\mu - 1)\,\tau]^{1-\frac{k}{\mu-1}}}{k - (\mu - 1)} \tag{21}$$

For the definition of $y(\xi, \tau)$ we shall use a correlation

$$\frac{\partial \gamma}{\partial \tau} = -\left(\frac{\partial \xi}{\partial \tau}\right)_{\gamma} \Big/ \left(\frac{\partial \xi}{\partial \gamma}\right)_{\tau} \tag{22}$$

and, from (2), (10) we get

$$y = -\frac{1}{(1-\gamma)^{\mu}} \cdot \left(\frac{\partial \xi}{\partial \tau}\right)_{\gamma} \Big/ \left(\frac{\partial \xi}{\partial \gamma}\right)_{\tau} \tag{23}$$

Also, from (21) follows that

$$\left(\frac{\partial \xi}{\partial \tau}\right)_{\gamma} = \frac{[1 + (\mu - 1)\,\tau]^{\frac{1}{\mu-1}}}{[1 + (\mu - 1)\,\tau]^{\frac{1}{\mu-1}} - 1}, \tag{24}$$

$$\left(\frac{\partial \xi}{\partial \gamma}\right)_{\tau} = 1/[\gamma(1-\gamma)^{\mu} \tag{25}$$

Introducing (24), (25) into (23) we get a correlation

$$y = \gamma \cdot \frac{[1 + (\mu - 1)\,\tau]^{\frac{1}{\mu-1}}}{[1 + (\mu - 1)\,\tau]^{\frac{1}{\mu-1}} - 1}. \tag{26}$$

The equations (21), (26) are the solution of the problem of the dynamics of sorption when the isotherm is intensely convex and the value of μ is taken arbitrarily. If μ is a whole number there arises a necessity for unfolding the uncertainty of the 0/0 type in one of the parts of the sum that corresponds to $k = (\mu - 1)$. If we assume that μ is not equal to but only approaches some whole number k and if we apply Lopitals rule, we get

$$\lim_{(\mu-1)\to k} \frac{(1-\gamma)^{k-(\mu-1)} - [1 + \tau(\mu - 1)]^{1-\frac{k}{\mu-1}}}{k - (\mu - 1)}$$
$$= \begin{cases} \ln(1-\gamma) + \tau & \text{when } \mu \to 1 \\ \ln(1-\gamma) + \dfrac{1}{\mu - 1} \ln[1 + \tau(\mu - 1)] & \text{when } \mu \to 2, 3 \ldots \end{cases} \tag{27}$$

$$\lim_{\mu\to 1} [1 + \tau(\mu - 1)]^{\frac{1}{\mu-1}} = e^{\tau} \tag{28}$$

If we employ the identity $\ln \frac{1}{1-x} = \sum_{k=1}^{\infty} \frac{x^k}{k}$ and take (27), (28), instead of (21), (26) then for a whole-number μ:

$$\xi = \tau + \ln \frac{1-\gamma}{\gamma} + \ln(1 - e^{-\tau}), \quad \mu = 1 \tag{29}$$

$$y = \gamma \frac{e^{\tau}}{e^{\tau} - 1}$$

and

$$\xi = \ln \frac{1-\gamma}{\gamma} + \ln \left\{[1 + (\mu - 1)\,\tau]^{\frac{1}{\mu-1}} - 1\right\} \tag{30}$$

$$+ \sum_{k=1}^{\infty} \frac{1}{k} [1 + (\mu - 1)\,\tau]^{\frac{1}{\mu-1}} - \sum_{k=1}^{\mu-1} \frac{1}{k(1-\gamma)^{\tau}}, \; \mu = 2, 3 \ldots \tag{31}$$

$$y = \gamma \frac{[1 + (\mu - 1)\,\tau]^{\frac{1}{\mu-k}}}{[1 + (\mu - 1)\,\tau]^{\frac{1}{\mu-1}} - 1} \tag{32}$$

From the solutions of (21), (26) we may obtain the asymptotic approximations. Let us rewrite (21) as follows

$$\xi = \sum_{k=0}^{\mu-1} \frac{(1-\gamma)^{k-(\mu-1)}}{k - (\mu - 1)} - \sum_{k=0}^{\infty} \frac{\left\{[1 + (\mu - 1)\,\tau]^{\frac{1}{\mu-1}}\right\}^{k-(\mu-1)}}{k - (\mu - 1)} \tag{33}$$

We can see that if $\mu > 1$ and $\tau \to \infty$ then the second sum in the right-hand part of (33) is a rapidly converging series. Then, confining ourselves to the first part we may approximately write for (33)

$$\tau = \xi - \sum_{k=0}^{\infty} \frac{(1-\gamma)^{k-(\mu-1)}}{k - (\mu - 1)} \tag{34}$$

or, in a dimensionless form (Shilov)

$$\tau = \xi - \xi_0, \; \xi_0 = \sum_{k=0}^{\infty} \frac{(1-\gamma)^{k-(\mu-1)}}{k - (\mu - 1)} \tag{35}$$

From (26) follows that if $\tau \to \infty$ then the correlation of Zeldowich is valid, that is $y \simeq \gamma$.

Fig. 1—3 show the development of the sorption process when $\mu = 1, 2, 3$. All stages of the process agree with Shilovs model [1]: the leap of concentrations when $\tau = 0 \left(t = \varepsilon \frac{x}{v}\right)$, determined by $y = \exp(-\xi)$; the formation of parallel transfer of the concentrations front; the advancement of the quasi-stationary front with the constant dimensionless rate. For the condition of parallel transfer Zeldovichs correlation is approximately valid, the accuracy

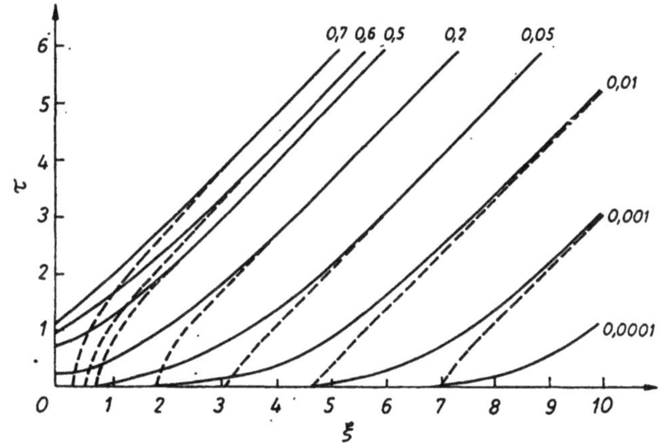

Fig. 1: Dynamics of adsorption in case of the sharply convex isotherm; $\mu = 1$.

– – – – - y = const ———- γ = const

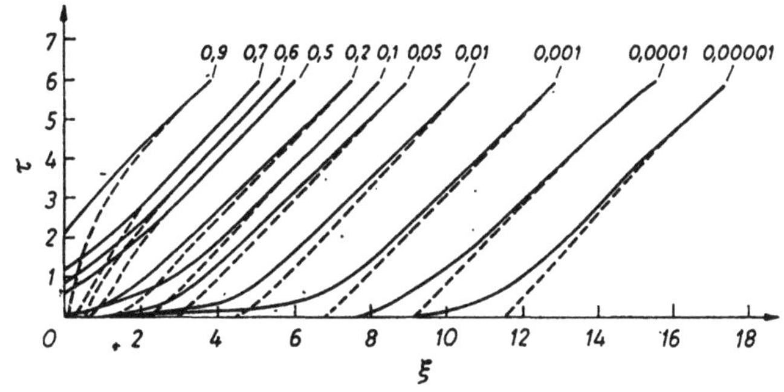

Fig. 2: Dynamics of adsorption in case of the sharply convex isotherm; $\mu = 2$.

– – – – - y = const ———- γ = const

of performance depending on τ and may be evaluated by a quantity

$$Z = \frac{y - \gamma}{y} = \frac{1}{[1 - (\mu - 1)\,\tau]^{\frac{1}{\mu - 1}}} \tag{36}$$

Analysis shows that when $\mu \to 1$, $Z = e^{-\tau}$, as when $\mu \to \infty$, $Z \to 1$ and with the rise in the value of μ the duration of formation of the quasi-stationary front grows also. For the regions where the concentrations y and γ are small, the set accuracy of the performance of the correlation $y \simeq \gamma$ is achieved sooner than for the regions higher concentrations. The duration of the formation of the quasi-stationary front, when μ is large, is prolonged considerably for the regions where the concentrations y approach 1. The situation may also arise, when for the explored range of the bed lengths the concentrations front in at mobile phase is moving under the condition of parallel transfer but the performance of the correlation $y \simeq \gamma$ is not accurate enough for the asymptotic approximation to be of practical use in engineering.

We may rewrite in dimensionless form the mathematical model (9) (8) in the following way:

$$\frac{\partial y}{\partial \xi} + \frac{\partial \gamma}{\partial \tau} = 0, \quad \frac{\partial \gamma}{\partial \tau} = y\left(1 - \frac{\gamma}{1+p}\right) - \frac{p}{1+p}\,y, \tag{37}$$

where, instead of $G = \frac{a}{a_\infty}$, above taken, $\gamma = \frac{a}{a_\infty}$.

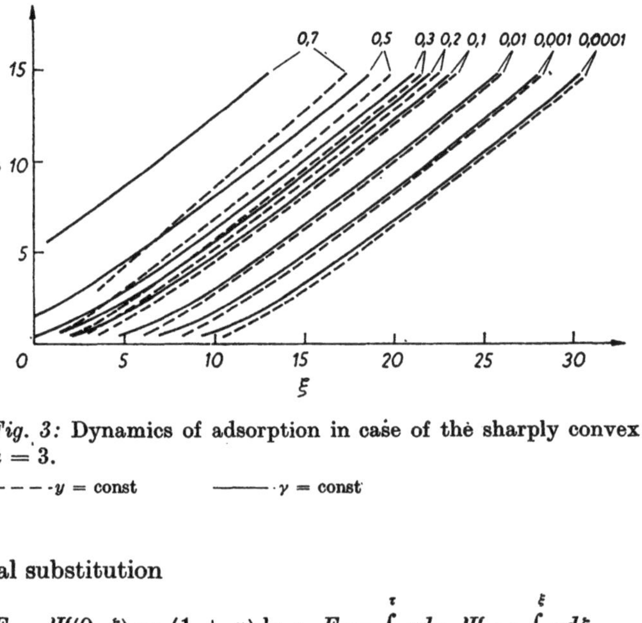

Fig. 3: Dynamics of adsorption in case of the sharply convex isotherm; $\mu = 3$.

- - - - y = const ——— γ = const

The integral substitution

$$F - \Psi(0, \xi) = (1 + p) \ln \varrho, \; F = \int_0^\tau y\,d\tau, \; \Psi = \int_0^\xi \gamma\,d\xi$$

makes possible the linearization of (37) and yields analytical expressions for the functions $y(\xi, \tau)$, $\gamma(\xi, \tau)$ by, e.g., an operative method. For the distribution of concentrations in a mobile phase it is convenient to have the solution in the following form:

$$y(\xi, \tau) = \frac{y_l\left(\tau, \frac{p}{1+p}\,\xi\right)}{y_l\left(\tau, \frac{p}{1+p}\,\xi\right) + e^{\frac{\xi-\tau}{1+p}}\left[1 - y_l\left(\frac{p}{1+p}\,\tau, \xi\right)\right]}, \tag{38}$$

where $y_l(s, q)$ is the solution of the problem of the dynamics of a frontal process when the isotherm is linear ($p \to \pm \infty$) that is

$$y_l(s, q) = e^{-(s+q)} \sum_{n=0}^{\infty} \frac{q^n}{n!} \sum_{m=n}^{\infty} \frac{s}{m!}. \tag{39}$$

The distribution of sorbate concentrations in an immobile phase $\gamma(\xi, \tau)$ can be obtained from a correlation that is easily verified:

$$\gamma_l(\tau, \xi) = 1 - \bar{y}_l(\xi, \tau) \tag{40}$$

where $y_e(\xi, \tau)$ corresponds to $y(\tau, \xi)$ with the arguments transposed.

Figures 4—6 show the development of the sorption process, described by the mathematical model (37).

For the qualitative analysis of the process it is obviously convenient to employ the solution in analytical form.

The properties of function (39) are rather completely established and it may be presented as a tabulated function. Considering the properties of function (39) we may establish the following laws:

A) If the isotherm is convex ($p > 0$) then for the region of concentrations grouped round the center of gravity of the sorption wave the following asymptotic dependence is correct

$$y = \frac{1}{1 + e^{\frac{\xi - \tau}{1+p}}}, \quad \tau \sim \xi \to \infty, \tag{41}$$

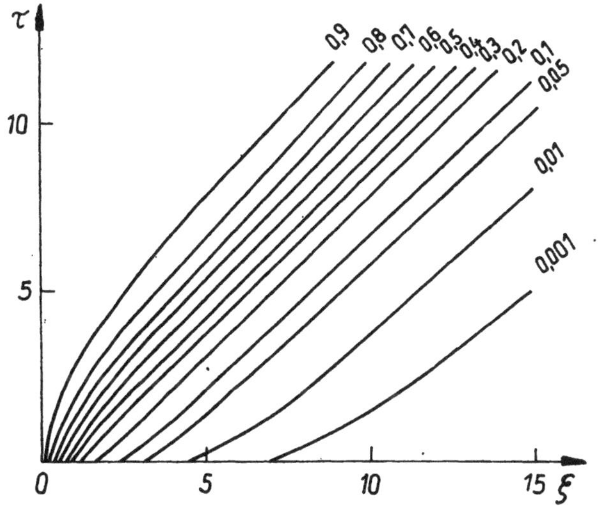

Fig. 4: Dynamics of adsorption for the case of linear isotherm

- - - - y = const ——— γ = const

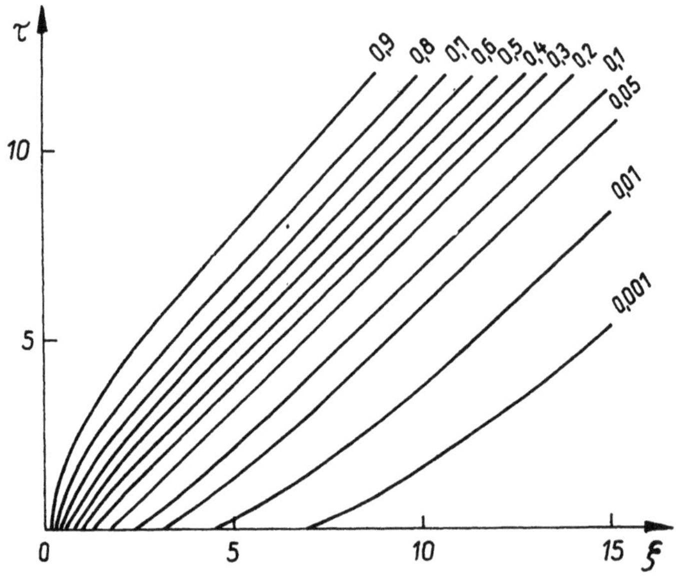

Fig. 5: Dynamics of adsorption for $P = 0.5$.
Kinetic equation (8) ———— $y = \text{const}$

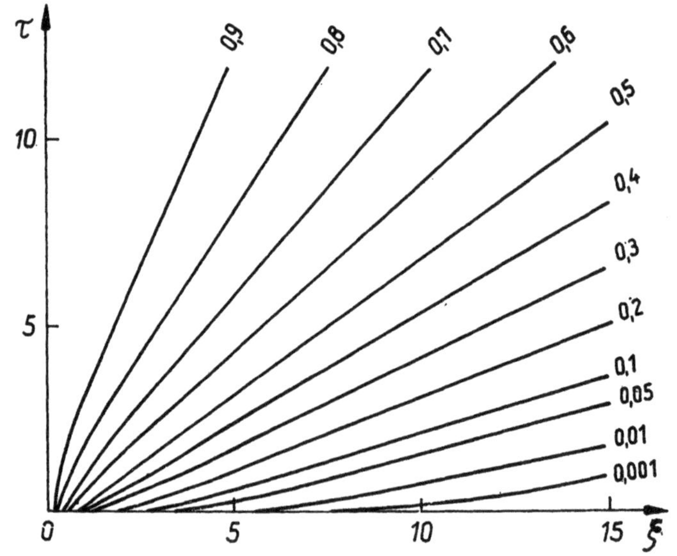

Fig. 6: Dynamics of adsorption for $p = -1.5$.
Kinetic equation (8) ———— $y = \text{const}$

For the region of small concentrations for any ξ, τ conforming to the conditions $\tau > \frac{p}{1+p}\,\xi$, $\xi > \tau$ the following expression is approximately correct

$$y \simeq e^{-\frac{\xi-\tau}{1+p}}, \quad \tau < \xi < \frac{1+p}{p}\,\tau. \tag{42}$$

Therefore, there is such an area of the sorption front, where the points of equal concentrations have practically from the very beginning of the process the same rate of movement, though the stationary front may be unformed yet. Observe, that the equation (41) may be rewritten also as a SHILOVS equation:

$$\tau = \xi - \tau_0, \ \tau_0 = (1+p)\ln\left(\frac{1}{y} - 1\right), \tag{43}$$

where τ_0 is a dimensionless waste of time of protective action.

In the region of high and medium concentrations (Fig. 5) the isopicns are convex at the initial stages. For the region where (42) becomes valid the isotherms are almost straight while when the concentrations are smaller the isopicns become concave. This occurrence has been frequently observed during the studies of the frontal process of absorption of solvent vapours by short beds of active coals for a wide range ($y \in [10^{-5}; 1]$) of relative "slip-through" concentrations [8] and it was commented upon by the authors of these publications as a certain characteristic property of the process.

B) If the isotherm is sharply convex ($p \to 0$) than the solution of a whole range of changes of ξ and τ assumes the following form:

$$y = \frac{1}{1 + e^{-\tau}(e^{\xi} - 1)}. \tag{44}$$

For the asymptotic stage the equation (44) assumes the form:

$$\tau = \xi - \tau_0, \ \tau_0 = \ln\left(\frac{1}{y} - 1\right) \tag{45}$$

also similar to SHILOVS equation.

C) When $p \to \pm\infty$ the sorption isotherm becomes linear and, from (38) $y = y_l(\xi, \tau)$. In this case there is no parallel transfer and the front of sorption widens in proportion to $\sqrt{\tau}$ when ξ and τ are large. The question of the asymptotic stage of the process is in need of further definition. Every asymptotic expression cannot represent an arbitrary form of an approximation to the accurate solution, there is only a limit of the approximation valid for some region of the vanables used as arguments. Naturally there should be no contradiction between the initial system of equations and its asymptotic solution. However, the existing practically usable asymptotic solutions by KLINKENBERG [9] and THOMAS [10] do not comply with the above-stated requirement when used for the regions where the difference between y and γ should be taken into account in order to maintain the necessary accuracy. Analysis shows, that the asymptotic solution of the problem of the dynamics of the frontal process, when the isotherm is linear, should be as follows:

$$y_l(\xi, \tau) = \frac{1}{2} \operatorname{erfc}\left(\sqrt{\xi} - \sqrt{\tau}\right) + \frac{1}{4\sqrt{\pi}\,(\xi \cdot \tau)^{1/4}}\, e^{-(\sqrt{\xi}-\sqrt{\tau})^2} \tag{46}$$

$$\gamma_l(\xi, \tau) = \frac{1}{2} \operatorname{erfc}\left(\sqrt{\xi} - \sqrt{\tau}\right) - \frac{1}{4\sqrt{\pi}\,(\xi \cdot \tau)^{1/4}}\, e^{-(\sqrt{\xi}-\sqrt{\tau})^2} \tag{47}$$

Also, the function $y_l + \gamma_l$ is symmetrical as to its center of gravity and the condition (40) is valid for all regions.

D) When $-1 > p > -\infty$ the isotherm becomes concave and the front of sorption is becoming indistinct while τ grows according to a dependence which is almost linear for the asymptotic stage. The expression (38), when ξ and τ are large; becomes indefinite as both the numerator and the denominator are coming to nought.

Unfolding the indefinity in (38) and employing the asymptotic expansions for Bessel's functions obtained by differentiating we get

$$y \simeq -p\left(1 - \sqrt{\frac{1+p}{p\tau} \cdot \xi}\right), \quad \frac{p}{1+p}\,\tau < \xi < \frac{1+p}{p}\,\tau. \tag{48}$$

Employing (46) we get from (48) an asymptotic expression for γ:

$$\gamma \simeq -(1+p)\left(\sqrt{\frac{p\tau}{(1+p)\,\xi}} - 1\right), \quad \frac{p}{1+p}\,\tau < \xi < \frac{1+p}{p}\,\tau. \tag{49}$$

If we combine (48) and (49), we obtain the correlation of concentrations in moving and fixed phases

$$\gamma \simeq \frac{(1+p) \cdot y}{p + y}. \tag{50}$$

Hence, at the asymptotic stage of the dynamic process and if the isotherm of sorption is concave then the distribution of sorbate over the phases is determined by the equation for the isotherm and the process approaches equilibrium. The numerical solution of the problem of the dynamics of sorption. For the general case of unlinear isotherm it is not possible yet to obtain an analytical solution of the problem for the stage of formation of the quasi-stationary front of concentrations.

For the numerical solution of the problems of the dynamics of sorption we used the method of nets [11, 12]. The net is composed of lines straight in ξ and τ directions. In order to obtain information on the functions of the net we used the most simple finite differences approximation for differential equationes

$$y_{i,j} - y_{i-1,j} = 0.5 \cdot \Delta\xi[(y - y^*)_{i,j} + (y - y^*)_{i-1,j}], \tag{51}$$

$$\gamma_{i,j} - \gamma_{i,j-1} = 0.5\Delta\tau[(y - y^*)_{i,j} + (y - y^*)_{i,j-1}]. \tag{52}$$

That this finite differences problem is convergent and stable is proved in [13].

Let us make a note of a special feature of the problem under consideration. For every layer $i = \text{const}$ the functions y and γ grow with τ in the considered frontal process and are equal to 1 only when $j \to \infty$. Taking into account the errors of calculations of R_y and R_γ, it is practical to calculate $y_{i,j}$ and $\gamma_{i,j}$ only up to the moment, when $|\gamma_{i,j} - 1| > R_\gamma$, otherwise it is expedient to assume $y_{i,j} = \gamma_{i,j} = 1$. The general laws for the formation of the quasi-stationary front of concentration depend, to a certain degree, upon the type of the kinetic equations and isotherms of sorption. Let us confine ourselves to a case of the sorption rate determined by the expression (5) and adsorption equilibrium described by the LANGMUIR equation. Then the mathematical model of the process in a dimensionless form will be as follows:

$$\frac{\partial y}{\partial \xi} + \frac{\partial \gamma}{\partial \tau} = 0, \quad \frac{\partial \gamma}{\partial \tau} = y - \frac{p\gamma}{1 + p - \gamma}. \tag{53}$$

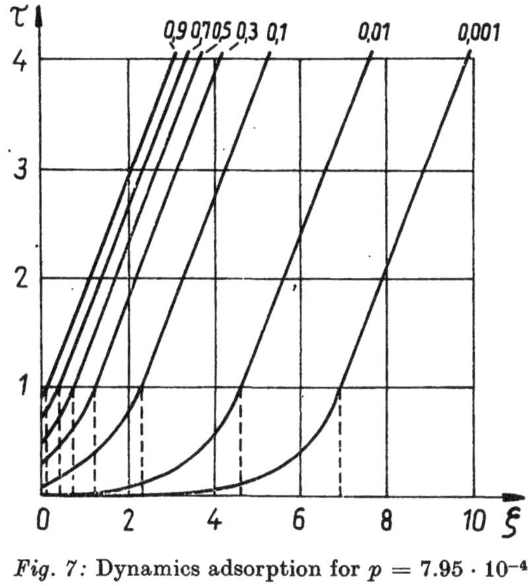

Fig. 7: Dynamics adsorption for $p = 7.95 \cdot 10^{-4}$
Kinetic equation (5)
- - - $y = \text{const}$; ——— $\gamma = \text{const}$

Figures 7—10 show some numerical solutions of (53). Analysis of the results obtained for the model shows that after the carrier flow has passed the cross-section of the bed that is under consideration and after the initial distribution of concentrations began to form behind the front of the flow, the point $y = \text{const}$ moves along the bed at a rate of U_y, which is less than the rate of a corresponding point $\gamma = y$. For small values of p there is for a time a constant slip-through concentration behind the bed. The higher the value of P the more rapid the movement of the point $y = \text{const}$ along the bed and the slower, the point $\gamma = \text{const}$. For convex isotherms ($p > 0$) the rates u_y and u_γ with the further development of the process tend to aquire constant values. This limit is the same for all of y, γ. Beginning with certain ξ_f, τ_f, that depend on p and y, for the inner region of the bed the following correlation is valid:

$$y(\xi, \tau) \simeq \gamma(\xi, \tau), \qquad u_y \simeq u_\gamma \simeq u_{ass}.$$

The values of ξ_f, τ_f, with which the sorption wave is formed with an accuracy not less than α, depend on p.

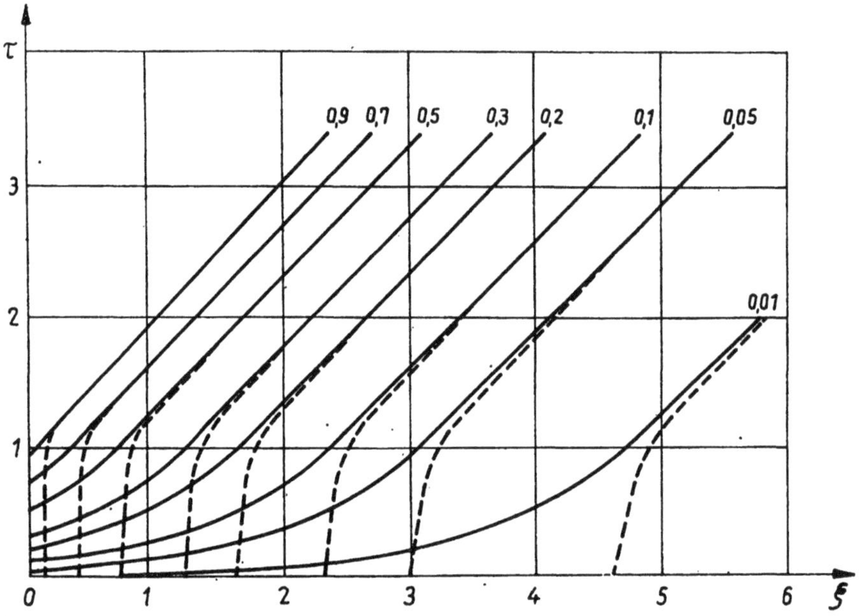

Fig. 8: Dynamics adsorption for $p = 0.05$.
Kinetic equation (5)
– – – -y = const; ——— -γ = const

In the area of small p, that is when the initial concentrations are high, or if the substances are readily adsorbed, or the sorbent has a sharply convex isotherm, the Zeldovich equation is rather accurate when $\tau_f \simeq 1$. In that case one may single out a special part of the exit curve — the constant "slip-through" concentration for some time behind the front.

Calculations showed that when $p \leq 10^{-2}$ there are no substantial differences in the dynamics of the process, which are stipulated by p itself.

The pictures show the development of the dynamics of the process for a wide range of relative concentrations y, γ.

The analysis of the results shows the concept of the quasi — stationary front to be applicable only to a limited range of concentrations. The less y and γ the slower the formation of the quasistationary front and the longer the zone

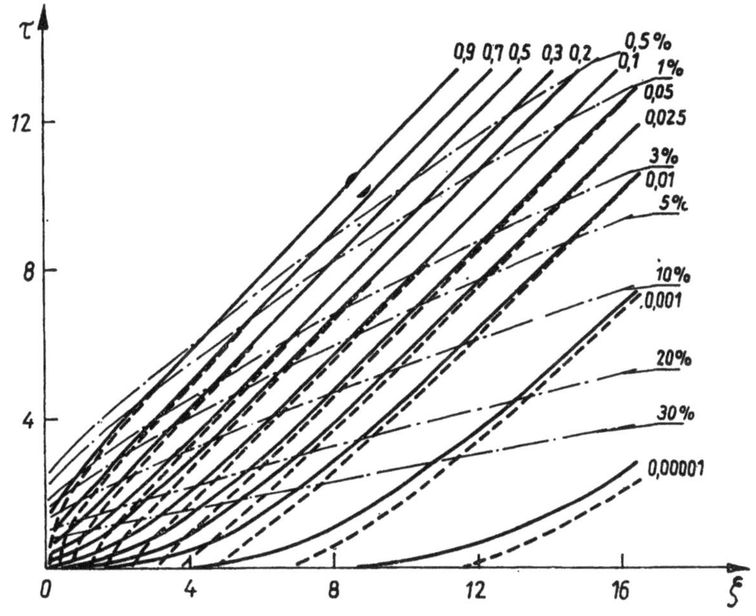

Fig. 9: Dynamics adsorption for $p = 0.5$.

Kinetic equation (5)

– – – – y = const ———— γ = const

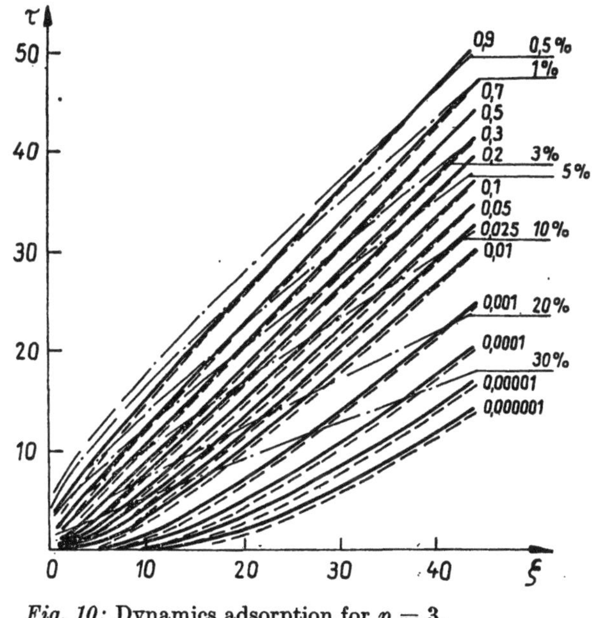

Fig. 10: Dynamics adsorption for $p = 3$.

Kinetic equation (5)

– – – – y = const ———— γ = const

of formation. With rise in p time τ_f and length ξ_f also rise necessary for the formation of the quasi-stationary conditions in a range $y_1 \div y_2$. $p \to \pm \infty$ corresponds to the linear isotherm and to the widening in time of the front of sorption.

For the frontal sorption processes, the rate of which is determined by equation (5), the formation of the quasi-stationary front for the $y_1 \div y_2$ range begins with higher y.

The result given above for the model show that the wide-spread idea that the formation of the quasi-stationary front ends with the practically complete exhaustion of the frontal layer of sorbent is correct only for sharply convex isotherms ($p \simeq 0$). It should be mentioned that for the carrier flow there may be established such points or even an area y that $U_y \simeq u_{ass}$ practically from the moment this concentration appears in the layer. It proves that not for all of the cases of movement of the point $y =$ const along the bed at a rate u_{ass} we may assume that throughout the bed there exists the quasi-stationary process. And first of all this applies to concentrations which are near or at the limit of sensitivity of the measuring instruments, because in this case it is impossible to study the movement of concentrations that are less than the abovementioned.

For the asymptotic stage of the process we may write SHILOV's equation:

$$\tau = \xi - \tau_0, \ \tau_0 = p \ln \left(1 - \frac{1}{y}\right) + \ln \frac{1}{y} - 1 \tag{54}$$

The system (6), (9) for the condition of parallel transfer of the concentrations front is solved in [14] for LANGMUIR and FREUNDLICH isotherms.

We may get the problem of the formation of the quasi-stationary concentrations front for the LANGMUIR isotherm from the solution of the system (5), (9) and through the following transformations:

$$y(\xi, \tau) = 1 - \tilde{\gamma}(\tau, \xi), \ \gamma(\xi, \tau) = 1 - \tilde{y}(\tau, \xi) \tag{55}$$

For the auxiliary functions $\tilde{\gamma}$, $\tilde{y}$ the system (5), (9) corresponds precisely to the system (4), (9) and it is obvious that its graphically presented solution may be used to form the correct ideas of the formation of the condition of parallel transfer when the kinetics are described by (5).

Fig. 11 gives an example of such transformation. In this case the momentary slip-through, formation of the quasi-stationary concentrations front and its parallel transfer along the bed are also easily traced. If for the frontal-type process with the kinetics described by the equation (5) the approximate equality of $y(\xi, \tau)$ to $\gamma(\xi, \tau)$ begins to be true first of all for the region of large y and γ then for the case, when the kinetics of adsorption corresponds to (6), the correlation $y(\xi, \tau) \simeq \gamma(\xi, \tau)$ is correct in the first place for the region of small y, γ. And also (Fig. 11) the area of low concentrations can move along the bed in the condition of parallel transfer beginning actually with the beginning of the process itself.

Naturally most processes take place having some intermediate conditions and the laws for the formation of the quasi-stationary front may be varied.

The approximate calculation of the influence of the longitudinal effects in the dynamics of adsorption is conditioned in the main part by the tendency to simplify the problem in such a way which makes it possible to obtain uncomplicated analytical dependencies at least for border-line cases. For numerical solutions of the problems the approximation achieved is not such a substantial one and a more detailed consideration of the problem is possible.

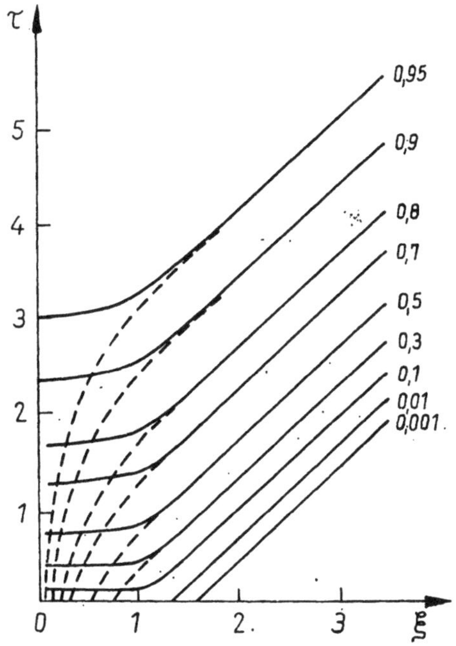

Fig. 11: Dynamics adsorption for $p = 0.05$.
Kinetic equation (6)
- - - - y = const ——— γ = const

In case of clear accounting for the influence of the effects of longitudinal transfer, the mathematical model for the frontal adsorption process can be written in the following way:

$$\frac{I}{G}\frac{\partial y}{\partial \tau} = D\frac{\partial^2 y}{\partial \xi^2} - \frac{\partial y}{\partial \xi} - \alpha\frac{\partial \gamma}{\partial \tau} \tag{56}$$

$$\frac{\partial \gamma}{\partial \tau} = y - \frac{p\gamma}{1 + p - \gamma} \tag{57}$$

$$y(\xi, 0) = \gamma(\xi, 0) = 0;$$

$$1 = y(0^+, \tau) - D\frac{\partial y(0^+, \tau)}{\partial \xi}; \quad \frac{\partial y(L^-, \tau)}{\partial \xi} = 0 \tag{58}$$

For the solution of this problem we likewise used the net method [...], employing the three-point non-obvious scheme. One example of the results of the calculation is given in Fig. 12. It becomes obvious, as we compare Fig. 8 and 12 that the influence of the longitudinal effects leads to some peculiarities as compared with the approximate calculations of them through k_{D_1} and k_{D_2}.

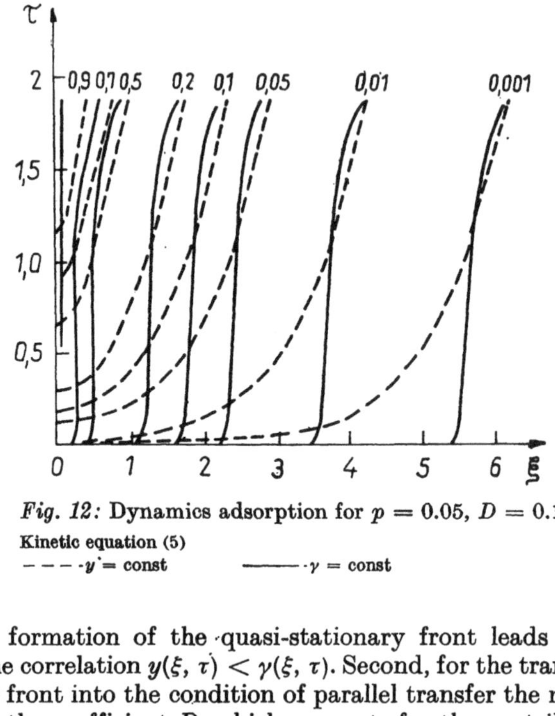

Fig. 12: Dynamics adsorption for $p = 0.05$, $D = 0.1$.

Kinetic equation (5)

- - - - $y = \text{const}$ —— $\gamma = \text{const}$

First, the formation of the quasi-stationary front leads at the asymptotic stage to the correlation $y(\xi, \tau) < \gamma(\xi, \tau)$. Second, for the transition of the quasi-stationary front into the condition of parallel transfer the more time is needed the bigger the coefficient D, which accounts for the contribution of the longitudinal effects in the washing-out of the concentrations front. As shown by the calculations and in comparison with numerous experimental data on adsorption of dissolved substances from liquids and vapours from air (adsorbents used — activated carbons, zeolites and oxides of aluminium), the clear accounting for the longitudinal effects is expedient only for the processes which are going rather rapidly and with low flow velocities, e.g. for the processes where fibrous adsorbents are used. In all the other cases an approximate accounting for the longitudinal effects is enough to assure accuracy needed for an engineering calculation.

References

[1] Shilov, N. A., Lepin, L. K., Vosnezenški, S. A., J. Russ. Phys. Chem. Soc. **61**, 1107 (1929)

[2] Rachinski, V. V., "*Introduction into General Theory of Adsorption Dynamics and Chromatography*", Nauka, Moscow 1962

[3] ZELDOWICH, Ja, B., Cit. according to [2]
[4] TREYBAL, R. F., "*Mass-Transfer Operation*", Mc Graw Hill, New York 1965
[5] TODES, O. M., BICSON, Ja. M, DAN USSR **75** (1950) 5, 725
[6] KOLIN, V. L., BAGROV, G. I., MUSAKIN, G. A., SEMENOV, V. P., "*Kinetics and Dynamics of Physical Adsorption*", Nauka, Moscow 1973
[7] LEVENSPIEL, O., BISCHOFF, K., Industr. and Engng. Chem. **51**, 12, 97 (1959)
[8] NIKOLAEV, K. M., POLYAKOV, N. S., DUBININ, M. M., "*Kinetics and Dynamics of Physical Adsorption*", Nauka, Moscow 1973
[9] KLINKENBERG, A., Ind. Eng. Chem. **40**, 1922 (1948)
[10] THOMAS, H. C., J. Am. Chem. Soc. **66**, 1964 (1944)
[11] WASOW, W. E., FORSYTHE, G. E., "*Finite Difference Methods for Partial Differential Equation*", Willy, New York 1960
[12] HAMMING, R. W., "*Numerical Methods for Scientists and Engineers*" Mc Graw Hill, New York 1962
[13] GODUNOV, S. K., RYABENKI, V. S., "*Introduction into the Theory of Difference Schemes*", Science, Moscow 1968
[14] GLUECKAUF, E., COATES, J. I., J. Chem. Soc., 1315 (1947)

Mixture adsorption dynamics

(A. A. Seballo)

Zusammenfassung:

Gegenstand des Beitrages sind mathematische Modelle der reversiblen und irreversiblen Adsorptionsdynamik von Mehrkomponentensystemen im Festbett.

An Beispielen der Adsorption von Zweikomponentengemischen und der Dynamik der Verdrängungsdesorption wird durch Simulation auf EDVA der Einfluß kinetischer Eigenschaften und Längsvermischungseffekte auf die Form der Ausgangskurven und der Verteilungskurven für einzelne Komponenten bei ihrer gleichzeitigen Adsorption oder gegenseitigen Verdrängung untersucht. Die Rechenergebnisse werden mit experimentellen Daten der Adsorption von 2 bis 4 Stoffgemischen verglichen.

Modern sorption processes, characterized by high intensity, represent, as a rule, separation or purification of the mixture of components or isolation of final components from the mixture of other components.

Adsorption dynamics regularities, based on fundamental concepts of adsorption equilibrium, kinetics and mass-energy transfer [1] are the foundation of the investigation, estimation and designing of sorption processes. At present these concepts are mostly developed for the asymptotic stage of the adsorption of single components. In conversion from adsorption dynamics of one components to adsorption dynamics of mixtures not only all the difficulties and unsolved problems are retained [1—3] but the number of variables increases sharply, which makes experimental investigations hardly realizable when there are more than two components. In this connection the necessity of the application of mathematical models and digital computers arose. Mathematical models used at present for describing mixture adsorption dynamics are classified according to the kinetic factors falling into two types: equilibrium and non-equilibrium [1, 4]. While working with equilibrium models it is supposed that the adsorption process proceeds indefinitely fast and the sorption dynamics of mixtures is characterized by the formation of stationary fronts according to the number of components. In case of the idealized model (lack of longitudinal expansion effects) steep fronts are formed moving along the layer in accordance with the sorption power of mixture components [1, 4].

In non-equilibrium models it is supposed that the expansion of sorption fronts results from the total influence of kinetic factors and longitudinal effects.

The quasistationary state having been established, expanded fronts moving with definite speed are generated along the length of the layer. The adsorption process of n-component mixture under equilibrium conditions proceeds in the following way.

Sorbate quantity UC_{i0}, which in section $x = 0$ is transferred by continuous flow $UC_i(0)$ and longitudinal transfer $-D_i^* \frac{dC_i(0)}{dx}$, is supplied into the column, which is H in length uniformly filled with pure adsorbent. In this case the mathematical model of the adsorption process is

$$\frac{\partial C_i}{\partial t} + \frac{\partial a_i}{\partial t} + U \frac{\partial C_i}{\partial x} = D_i^* \frac{\partial^2 C_i}{\partial x^2} \tag{1}$$

$$a_i = f_i(c_1, c_2, \dots, c_n) \qquad i = 1, 2, \dots n$$

$$t = 0 \quad H \geq x \geq 0 \qquad C_i = a_i = 0 \tag{2}$$

$$t > 0 \begin{cases} x = 0 & UC_{0i} = UC_i(0) - D_i^* \frac{dC_i(0)}{dx} \\ x = H & D_i^* \frac{dC_i}{dx} = 0 \end{cases} \tag{3}$$

The solution of this system for non-linear isotherm permits to obtain the mathematical expression for the speed of the stationary front of single components [1]. The shape of component fronts and the influence of the process parameters on the former can be determined only by numerical integration.

When non-equilibrium process is considered the mathematical model of mixture adsorption becomes

$$\frac{\partial C_i}{\partial t} + \frac{\partial a_i}{\partial t} = -u \frac{\partial C_i}{\partial x} \tag{4}$$

$$\frac{\partial a_i}{\partial t} = \beta_i[C_i - \varphi_i(a_1 a_2, \dots a_n)] \qquad i = 1, 2, \dots n$$

$$t = 0 \qquad C_i(x, 0) = a_i(x, 0) = 0 \tag{5}$$

$$t > 0 \qquad x = 0; \qquad C_i(0, t) = C_{0i} \tag{6}$$

The solution of this system for binary and three-component mixtures yields an expression for the speed of component stationary fronts and maximal values of displaced component concentrations. In order to estimate front concentrations and recognize their dependences from kinetic characteristics of single component it is necessary to use numerical methods.

For the practical application of previously mentioned models for digital computers it is necessary to define β_i, D_i^* and the dependence of equilibrium concentrations of the components during their adsorption in mixture with other components on adsorption values, i.e. $C_i^* = \varphi_i(a_1, a_2 \dots, a_n)$. In case of an analytical expression of mixture adsorption isotherms equilibrium concentrations can be determined readily. However, mostly this dependences are unknown, which makes the estimations hardly realizable. At the same time the experimental determination of equilibrium concentrations even for several components requires the carrying out of so many labour-consuming experiments that their duration makes further calculations on digital computer

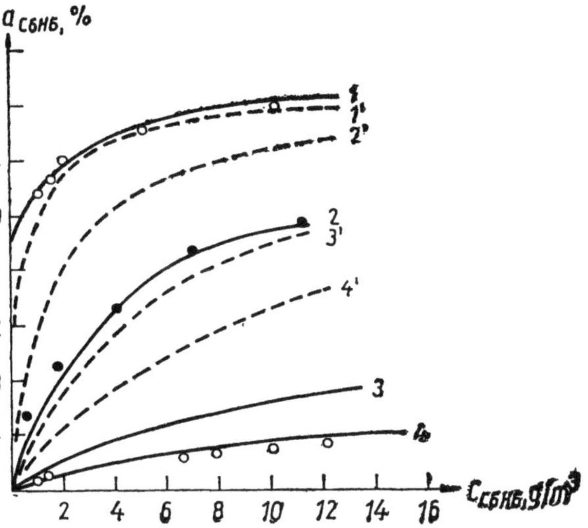

Fig. 1: Adsorption isotherm for benzene on SKT at $C_{C_8H_{18}}$ values
1 – 0 g/m³; *2* – 1 g/m³; *3* – 5 g/m³; *4* – 10 g/m³.

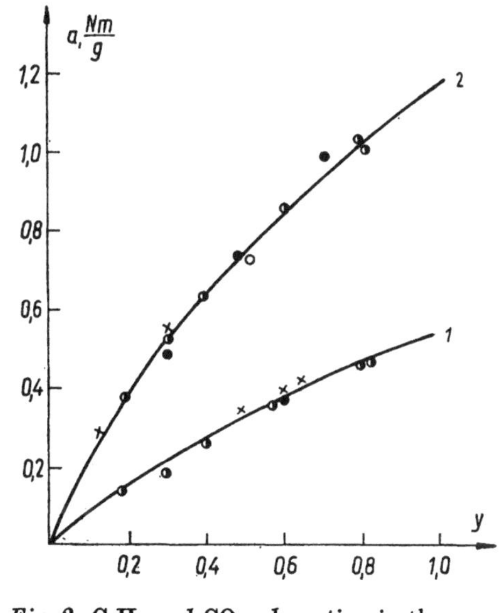

Fig. 2: C_2H_2 and CO_2 adsorption isotherm.
1 – P = 50 mm Hg; *2* – P = 250 mm Hg
◑ – theory
● – LANGMUIR
× – MYERS-PRAUSNITZ

unreasonable. In order to obtain equilibrium concentrations various methods were used. Fig. 3 shows an experimental dynamic adsorption isotherm of benzene-n-octan mixture on activated coal in the form of dependence $a_1 = f(C_1)$ $c_2 = \text{const}$.

The calculation of the mixture results from the individual isotherm parameters, the determination of which was accomplished conventionally, i.e. by the reduction of LANGMUIR equation to the linear type and determination of $1/C$ and $1/a$ by the tangens of the straight line angle of inclination to the ordinate cut by this line. Isotherms determined in such a way differ significantly

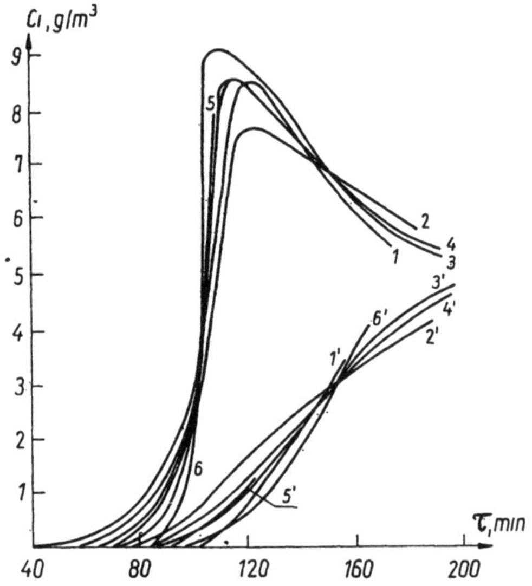

Fig. 3: Breakthrough curves for binary mixture adsorption.
1,1' – $\beta_1 = 30$, $\beta_2 = 10$; 2,2' – $\beta_1 = 10$, $\beta_2 = 10$;
3,3' – $\beta_1 = 20$, $\beta_2 = 15$; 4,4' – $\beta_1 = 15$, $\beta_2 = 15$;
5,5' – $\beta_1 = 20$, $\beta_2 = 20$; 6,6' – $\beta_1 = 30$, $\beta_2 = 30$ 1/s.
$u = 28{,}1$ cm/s; $H = 20$ cm; $c_1 = c_2 = 5$ g/m³.

from the experimental ones, the adsorption value of less sorbing benzene being over-rated and of n-octane on the contrary being underrated. In order to decrease deviation between estimated and experimental data any minimizing method can be used. In case of binary mixture adsorption we arrived at the problem of minimizing the following expression

$$F = \sum_j \sum_i \left(a^j_{1i} - \frac{A_1 \Gamma_1 C^j_{1i}}{A_1 + \Gamma_1 C^j_{1i} + \Gamma_2 C^0_{2j}}\right)^2 + \\ + \sum_i \sum_j \left(a^i_{2j} - \frac{A_2 \Gamma_2 C^i_{2j}}{A_2 + \Gamma_1 C^i_{1j} + \Gamma_2 C^j_{2i}}\right)^2 \qquad (7)$$

for all values A_1, A_2, Γ_1, $\Gamma_2 \geqq 0$

Minimum localization was accomplished with the help of a digital computer by the method treated in [5]. After parameter determination it is possible to accomplished calculations of any combinations of C_1 and C_2. Comparing calculated results with experimental data showed that the application of the given approach permits to attain satisfactory coincidence of calculated and experimental isotherms. Evidently, for mixture adsorption isotherm calculation it is necessary in this case to have not less than 3 curves of $a_1 = f_1(C_1)_{C_2=\text{const.}}$ and $a_2 = f_2(C_2)_{C_1=\text{const.}}$ forms, experimental determination of which is a complex and labour-consuming process connected moreover with considerable errors. For the purpose of the calculation of adsorption isotherm parameters of single components the method of least squares can be used. In this case initial data are defined in the form of individual adsorption isotherm (table $a_i = f(C_i)$). Then A_i and Γ_i are determined and used for the following determination of mixture adsorption isotherm and equilibrium concentrations for any combination of the concentrations of these substances.

In order to develop new methods of the estimation of equilibrium concentrations an attempt to use the MYERS-PRAUSNITZ approach based on the concept of ideal solutions was made [6]. As the method is a graphical one it introduces significant errors in determining equilibrium concentrations in the very important field — where there is slope rise of the adsorption isotherm. Moreover, values of adsorption determined in such a way are practically unusable in digital computers due to lack of convenient analytical dependence $C_i = \varphi_i \times (a_1, a_2 \ldots a_n)$ used in mixture adsorption dynamics estimation. Methods of digital computer calculation of mixture adsorption isotherms and equilibrium concentrations with various types of equations for describing adsorption statics of single component are stated below.

According to [6] for the mixture we have

$$\Pi A = (\Pi A)_1 + (\Pi A)_2 + \ldots + (\Pi A)_n \tag{8}$$

As the individual isotherms were defined on the same adsorbent as the mixture adsorption isotherm the values of A were in all experiments the same

$$\Pi A = RT \int_0^{p_i} \frac{a_i}{p_i}\, dp_i . \tag{9}$$

The relationship of values p_{01} and p_{02} can be obtained by equating to each other expressions for two-dimensional pressures along the individual isotherms

$$RT \int_0^{p_1^0} \frac{a_1^0}{p_1^0}\, dp_1^0 = RT \int_0^{p_2^0} \frac{a_2^0}{p_2^0}\, dp_2^0 = \cdots = RT \int_0^{p_n^0} \frac{a_n^0}{p_n^0}\, dp_n^0 \tag{10}$$

As $p \sim c$, it follows from (10) that

$$\int_0^{c_1^0} \frac{a_1^0}{c_1^0}\, dc_1^0 = \int_0^{c_2^0} \frac{a_2^0}{c_2^0}\, dc_2^0 = \cdots = \int_0^{c_n^0} \frac{a_n^0}{c_n^0}\, dc_n^0 \tag{11}$$

If individual isotherms are approximated by the equation of LANGMUIR type

$$a_i = \frac{A_i \Gamma_i C_i}{A_i + \Gamma_i C_i}; \qquad b_i = \frac{\Gamma_i}{A_i} \tag{12}$$

then expression (11) with allowance for (8) acquires the following form:

$$(\Pi A)_i^0 = RT \int_0^{c_i^0} \frac{a_i^0}{c_i^0}\, dc_i^0 = RT \int^{c_i^0} \frac{A_i b_i}{1 + b_i c_i}\, dc_i^0$$

$$= RT \ln\,(1 + b_i c_i^0), \qquad c_i = \frac{(1 + b_1 c_1^0)^{A_1/A_2} - 1}{b_i}. \tag{13}$$

When C_1^0 is determined on the basis of the relations of (6) and when its values are substituted into the expression $x_1 + x_2 + \cdots x_n = 1$.
The following equation is obtained:

$$\frac{c_1}{c_1^0} + \sum_{i=1}^{i=n} \frac{c_i b_i}{(1 + b_i c_i^0)^{A_1/A_2} - 1} = 1 \tag{14}$$

C_1^0 can be found from non-linear equation (14) C_i^0 from (13) and x_i from (14).

The total amount of the adsorbed substance in accordance with [6] is determined from

$$\frac{1}{a_0} = \frac{x_1}{a_1^0} + \frac{x^2}{a_2^0} + \cdots \frac{x_n}{a_n^0} \tag{15}$$

where a_i^0 – amount of ith component adsorbed under superficial pressure Π and lack of other components; a_i^0 values are determined from the individual adsorption isotherm by c_i^0 values.
Then

$$\frac{1}{a} = \sum_{i=1}^{i=n} \frac{x_i(1 + b_i c_i^0)}{A_i b_i c_i^0} \tag{16}$$

Knowing the total amount of adsorbed substance a_{total} it is possible to obtain the adsorption values of the single components. Thus the estimation of the mixture adsorption isotherm, i.e. the determination of dependences is accomplished in the following way. For simplicity of treatment let us consider the estimation of binary mixture adsorption isotherms. Parameters A_1, A_2, b_1, b_2, are determined with a digital computer by the least squares method with the single component adsorption isotherm. Then c_1 is calculated from eq. (14), and from (15) a_1 and a_2 are determined. In estimating equilibrium concentrations as well as in the direct problem of adsorption value estimation with known concentrations A_1, A_2, b_1, b_2 are defined. Simple transformation gives

$$\frac{x_1(1 + b_1 c_1^0)}{A_1 b_1 c_1^0} + \frac{x_2(1 + b_1 c_1^0)^{A_2/A_1}}{A_2(1 + b_1 c_1^0)^{A_1/A_2} - 1} = \frac{1}{a_1 + a_2}. \tag{17}$$

Solving (17) we obtain C_1^0 and then C_1 and C_2. In order to plot the values of $c_i = \varphi_i(a_1, a_2)$ it is necessary to define change limits of a_1 and a_2 and the step of this change. As the function increases sharply under low values of a_1 the step of change of Δa_1 should be small.

Adsorption isotherm estimation results for the mixture of ethylene and carbon dioxide [7] are shown in Fig. 2. Figure 2 shows results of estimation by MYERS-PRAUSNITZ graphical method, the LANGMUIR equation parameters of which were defined by a digital computer. As it is seen from these data the suggested method permits with accuracy sufficient for practical purposes to estimate mixture adsorption isotherms.

Approximating experimental isotherms by KISAROV's equation gives

$$(\Pi A)_i^0 = RT \int_0^{c_i^0} \frac{a_i^0}{c_i^0}\, dc_i^0 = RT \int_0^{c_i^0} \frac{A_i b_i (c_i^0/c_{si})^{n_i}}{(1 + b_i c_i^0/c_{si})^{n_i} \cdot c_i}\, dc_i^0 \tag{18}$$

Simple transformation gives definitely

$$\frac{c_1}{c_1^0} + \frac{c_2 b_2^{1/n_2}}{\left\{[1 + b_1(c_i^0/c_{s1})^{n_1}]\,\dfrac{A_1 n_2}{A_2 n_1} - 1\right\}^{1/n_2} \cdot c_{s2}} - 1 = 0, \tag{19}$$

the solution of which permits to find c_i^0 and then equilibrium concentrations and adsorption values of mixture components. In case when individual isotherms are approximated poorly by the above equations but rather well by DUBININ-RADUSKEVITCH's or DUBININ-ASTACHOV's equations we have

$$a_i^0 = \frac{W_0}{v_i^0} \exp\left[-\frac{BT^2}{\beta_{0i}^2}\left(\lg \frac{c_{si}}{c_i^0}\right)^2\right]; \qquad a_i^0 = a_0 \exp\left[-\left(\frac{A}{E}\right)^n\right]. \tag{20}$$

Then

$$(\Pi A)_i^0 = RT \int_0^{c_i^0} \frac{a_i^0}{c_i^0}\, dc_i^0 = RT \int_0^{c_i^0} \frac{c^0 W_0}{c_i^0 v_i^0} \exp\left[-\frac{BT^2}{\beta_{0i}^2}\left(\lg \frac{c_{si}}{c_i^0}\right)^2\right] dc_i^0. \tag{21}$$

The relationship of c_i^0 and c_2^0 can be found by equating to each other equations of two-dimensional pressures along individual isotherm lines

$$\int_0^{c_i^0} \frac{a_1^0}{c_1^0}\, dc_1^0 = \int_0^{c_2^0} \frac{a_2^0}{c_2^0}\, dc_2^0 \tag{22}$$

It follows from (14)—(15) that

$$c_2^0 = \frac{c_1^0 c_2}{c_1^0 - c_1}. \tag{23}$$

Then for the determination of c_1^0 we obtain from (23) and (22)

$$\int_0^{c_1^0} \frac{a_1^0}{c_1^0} - \int_0 \frac{c_1^0 c_2}{c_1^0 - c_1} \cdot f(c_1^0)\, dc_1^0 = \varphi(c_1^0) \tag{24}$$

Thus the problem is reduced to root determination function $\varphi(c_1^0)$ where

$$f(c_1^0) = a_0 \exp\left[-\left(\frac{RT}{E} \ln \frac{c_s(c_1^0 - c_1)}{c_1^0 c_2}\right)^n\right] \cdot \frac{c_1}{c_1^0(c_1 - c_1^0)} . \tag{25}$$

This approach is readily seen to expand to a larger number of components.

Initial data for estimation are individual component characteristics only: W_0, B, β_1, β_2, V_1, V_2, T, C_{s1}, C_{s2} or a_0, E, A of equation (20). Table 1 shows the results of estimation by a digital computer of $C_2H_5Cl-(C_2H_5)O$ binary mixture adsorption isotherms determined by BERING and SERPINSKY [7].

Table 1: Adsorption isotherms for $(C_2H_5)O$ and $(C_2H_5)Cl$

Experiment					Theory			
P_1	P_2	$P_{1,2}$	$a_{1,2}$	x_2	$a_{1,2}$	x_2	$\frac{\Delta a}{a_{exp.}}$	$\frac{\Delta x_2}{x_{2exp.}}$
17.1	36.9	54.0	3.75	0.870	3.75	0.912	0	4.8
17.8	12.9	30.7	3.54	0.794	3.58	0.824	1.1	3.8
31.5	104.6	136.1	4.02	0.861	4.04	0:921	1.2	7.0
32.5	14.2	46.7	3.72	0.737	3.73	0.726	0.3	1.3
49.3	18.1	67.4	3.90	0.675	3.92	0.668	0.5	1.0
68.0	21.8	89.8	4.07	0.625	4.08	0.618	0.2	1.1
70.2	50.7	120.9	4.19	0.690	4.16	0.749	0.7	7.8
83.0	115.0	198.0	4.25	0.735	4.24	0.820	0.2	11.5
118.6	31.6	150.2	4.34	0.519	4.40	0.534	0.5	2.9
149.0	38.4	187.2	4.49	0.450	4.43	0.507	1.3	11.4
274.5	33.9	308.4	4.87	0.323	5.00	0.302	2.7	6.2
							$\frac{\Delta a}{a} \sim 0{,}9\%$	$\frac{\Delta x_2}{x_2} \sim 5{,}3\%$

It is seen from the table data that the mean deviation of the estimation value a from the experimental ones is 0.9% while the mean deviation of mole fraction x_i is 5.3%, which is suitable for estimations. As only individual component characteristics are necessary for estimations they can be accomplished when adsorbent structural characteristics defined according to standard substance for various relationships of concentrations, temperatures etc. are given.

In the presence of single component adsorption experimental isotherms defined by the table of $a_i = f(c_i)$ it is possible to approximate them by polynomials of the following type

$$y = a_0 + a_1 x + a_2 x^2 + a_3 x^3 + \cdots + a_n DC^n \qquad (26)$$

In this case the problem of adsorption isotherm estimation is reduced to the following. Analogously to (22)

$$\int_0^{c_{01}} a_0 + a_1 c_{01} + a_2 c_{01}^2 + \cdots = \int_0^{c_{02}} b_0 + b_1 c_{02} + b_2 c_{02}^2 + \cdots \qquad (27)$$

Then from (27) and (14) we obtain the expression for the determination of C_{01}

$$\int_0^{c_{01}} a_0 + a_1 c_{01} + \cdots = \int_0^{\frac{c_{01} c_2}{c_{01} - c_1}} b_0 + b_1 \frac{c_{01} \cdot c_2}{c_{01} - c_1} + \cdots \qquad (28)$$

The estimation of equilibrium concentrations and mixture adsorption values is similar to that mentioned above. Thus for the estimation of mixture adsorption isotherms and equilibrium concentrations used in mixture dynamics all characteristics are defined in the form of single componente adsorption isotherm parameters A_i, b_i; W_0, B; A, E, n. In case when mixture adsorption isotherms cannot be represented in analytical form and experimental data are meagre it is possible to define them in the form of tabular values plotted by various methods including all those mentioned above.

The choice of particular values of $c_i^* = f(a_1, a_2, \ldots a_n)$ from the table is accomplished after interpolation by polynomials of the second and third power.

In order to estimate kinetic characteristics of (4) various methods are applied including method [8] based on mass-exchange coefficient determination by employing a digital computer.

On account of the lack of experimental methods for $\beta_i = f_i(a_1, a_2 \ldots a_n)$ for the solution of system (4) average integral values of the single components of the transfer coefficients are used. The estimation of D_i^* values can be carried out approximately by the LEVEHSPIEL equation [9]

$$D_i^* \approx \frac{1}{2} d \cdot u. \qquad (29)$$

In general the effects connected with longitudinal diffusion are similar to [3] effects, caused by external diffusion with the general mass-transfer coefficient and can be determined by employing the formula

$$D^* = \frac{v^2}{\beta_e} \qquad (30)$$

where

$$\frac{1}{\beta_e} = \frac{1}{\beta_{out}} + \frac{1}{\beta_{in}}$$

In order to solve system (1) it is necessary to reduce it to the dimensionless form

$$\mu \frac{\partial y_i}{\partial \tau} + \mu \frac{A_i}{C_{0i}} \frac{\partial Y_i}{\partial \tau} + \frac{\partial y_i}{\partial \xi} = G_i \frac{\partial^2 c_i}{\partial \xi^2} \tag{31}$$

$$Y_i = \frac{y_i}{p_i \left(1 + \sum_{m=1}^{n} \frac{y_m}{p_m} \frac{A_m}{A_i}\right)}$$

where

$$\xi = \frac{x}{H}; \quad \tau = \frac{ut \cdot c_{01}}{H \cdot A_1}; \quad y_i = \frac{c_i}{c_{0i}}; \quad Y_i = \frac{a_i}{A_i};$$

$$\mu = \frac{c_{01}}{A_1}; \quad G_i = \frac{D_i}{H \cdot U}; \quad p_i = \frac{A_i}{\Gamma_i c_{0i}}.$$

Initial and boundary conditions are

$$\tau = 0; \qquad 1 \leq \xi \leq 0; \qquad y_i = Y_i = 0$$

$$\tau > 0 \begin{cases} \xi = 0; \quad 1 = y_i(0) - G_i \dfrac{dy_i(0)}{d\xi} \\ \xi = 1; \quad G_i \dfrac{dy_i}{d\xi} = 0 \end{cases}$$

For solving this problem it is more reasonable to apply explicit schemes which are completely stable under any step. As Y_i acquires the following form (31) we can write

$$Y_i = \frac{y_i \varphi_i}{p_i}; \quad \varphi_i = \frac{1}{\left(1 + \frac{y_1}{p_1} \frac{A_1}{A_i} + \frac{y_2}{p_2} \frac{A_2}{A_i} + \cdots + \frac{y_n}{p_n} \frac{A_n}{A_i}\right)}$$

System (31) was divided into n equations. For the solving of ith equation double-layer numerical system

$$\mu \frac{y_{i,j,k+1}^{(s+1)} - y_{i,j,k}}{l} +$$

$$+ \mu \Gamma_i \left(\frac{y_{i,j,k+1}^{(s+1)} - y_{i,j,k}}{l} \varphi_{i,j,k} + y_{i,j,k+1}^{(s+1)} \frac{\varphi_{i,j,+1}^{(s)} - \varphi_{i,j,k}}{l} \right) +$$

$$+ \frac{y_{i,j,k+1}^{(s+1)} - y_{i,j,k+1}^{(s+1)}}{h} = G \frac{y_{i,j+1,k+1}^{(s+1)} - 2y_{i,j,k+1}^{(s+1)} + y_{i,j+1,k+1}^{(s+1)}}{h^2}$$

where $y_{i,j,k+1}^{(s+1)} - y_{i,j,k+1}$ function iteration $(s + 1)$ value

$$\varphi_{i,j,k+1}^{(s)} = f_{i,j}(y_{1,j,k+1}^{(s)}, y_{2,j,k+1}^{(s)} \cdots y_{n,j,k+1}^{(5)})$$

was used.

By way of initial approximation values of function at the preceding step of time were selected.

In the calculation with the help of an iteration scheme either the number of iterations or the accuracy were defined and fulfilment of the following conditions was required

$$j^{\max} i^{\max} \,|\, y_{i,j,k+1}^{(s+1)} - y_{j,i,k+1}^{(s)} \,| \leqq \varepsilon .$$

The drawback of the system is that iteration calculation requires doubling of the number of memory cells occupied in the computer but this scheme permits to use a large-scale time step which results in the diminishing of calculation operations. As a result of calculations concentration distributions of all the components both in solid and in flowing (gas, liquid) phases along the layer length in time and their output curves beyond the adsorbed layer are estimated. The solving of system (4) is accomplished by a digital computer after the system has been reduced to dimensionless form.

$$\frac{C_{01}}{a_{01}} \frac{\partial y_1}{\partial \tau} + \frac{\partial y_1}{\partial \tau} = -\frac{\partial y_1}{\partial \xi}; \quad \varkappa \frac{c_{01}}{a_{01}} \frac{\partial y_2}{\partial \tau} + \frac{\partial Y_2}{\partial \tau} = -\varkappa \frac{\partial y_2}{\partial \xi}; \tag{32}$$

$$\frac{\partial Y_1}{\partial \tau} = y_1 - \varphi_1(Y_1, Y_2); \quad \frac{\partial Y_2}{\partial \tau} = \psi_i \varkappa [y_2 - \varphi_2(Y_1, Y_2)]$$

where

$$\xi = \frac{x\beta_1}{u}; \quad \tau = \frac{t \cdot c_{01} \cdot \beta_1}{a_{01}}: \quad y_i = \frac{c_i}{c_{0i}};$$

$$L = \frac{H\beta_1}{u}; \quad Y_i = \frac{a_i}{a_{0i}}; \quad \psi_i = \frac{\beta_i}{\beta_1}; \quad \varkappa = \frac{c_{02} \cdot a_{01}}{a_{02} \cdot c_{01}};$$

Taking into consideration that $c_{01}/a_{01} \approx 10^{-3} \div 10^{-6}$, value $\varkappa$ is close to 1, the change of relative concentration in time as a result of large adsorption value is negligible and it is possible to neglect terms including $\frac{\partial y}{\partial \tau}$ in the left parts of (32).

In this case system (32) acquires the following form

$$\frac{\partial Y_1}{\partial \tau} + \frac{\partial y_1}{\partial \xi} = 0; \quad \frac{\partial Y_2}{\partial \tau} + \varkappa \frac{\partial y_2}{\partial \xi} = 0 \tag{33}$$

In (33) y_i is continuous for all values of ξ and τ, except when $x = 0$ at $t = 0$, as relative concentrations of components increase from 0 to 1. In order to smooth this break in the solution of this problem let us begin at time t' i.e. $t' = t - x/u$. The substitution of (32) into (33) and taking into consideration that at the initial moment of time $\tau = 0$, $Y_i = 0$ we obtain

$$\frac{dy_1}{d\xi} = y_1; \quad \frac{dy_2}{d\xi} = -\psi y_2$$

with the boundary conditions

$$\xi = 0 \qquad y_i = 1$$

Solution

$$y_1 = e^{-\xi} \qquad y_2 = e^{-\varphi_i \xi}$$

gives initial distributions of the relative concentrations of components in the moving (flowing) phase. Then initial and boundary conditions of the problem will be of the following form:

$$\tau = 0 \qquad Y_i(\xi, 0) = 0; \qquad y_i(\xi, 0) = e^{-\varphi_i \xi} \tag{34}$$

$$\tau > 0 \qquad \xi = 0; \quad y_i = 1, \qquad Y_i(0, \tau) = f(\tau) \tag{35}$$

Values of Y_i $(0, \tau)$ are determined from the solution of the system when

$$\frac{dY_i}{\partial \tau} = \psi_i x[1 - \varphi_i(Y_1, Y_2)] \qquad i = 1, 2. \tag{36}$$

$$\tau = 0, \ Y_i = 0. \tag{37}$$

In order to solve (32) under the conditions of (34)—(35) let us use the numerical method. We designate the steps along the axes ξ and τ by h and l. We want to express derivatives by values of functions in the grid nodes with an accuracy defined by the differential quotient. Let us calculate values of functions Y_i, y_i appearing in the equations (32) by their values at the neighbouring points. In calculating Y_i, y_i at point $k + 1$, $j + 1$ values Y_i and y are assumed to be equal

$$Y_i = \frac{Y_{i,k+1,j+1} + Y_{i,k+1,j}}{2}; \qquad y_i = \frac{y_{i,k+1,j+1} + y_{i,k+1,j}}{2} \tag{38}$$

and in calculating $y_{i,k+1,j+1}$ they are:

$$Y_i = \frac{Y_{i,k+1,j+1} + Y_{i,k,j+1}}{2}; \qquad y_i = \frac{y_{i,k+1,j+1} + y_{i,k+1,j}}{2} \tag{39}$$

Substituting expressions (38), (39) into (32) we obtain

$$Y_i = \frac{Y_{i,k+1,j+1} + Y_{i,k+1}}{2} = \psi_i x \left\{ \frac{y_{i,k+1,j+1} + y_{i,k+1,j}}{2} \right.$$

$$\left. - \frac{1}{2} \left[\varphi_i(Y_{1,k+1,j+1} Y_{2,k+1,j+1}) + \varphi_i(Y_{1,k+1,j} Y_{2,k+1,j}) \right] \right\} \tag{40}$$

$$Y_{i,k+1,j+1} = Y_{i,k+1,j} + R_i(y_{i,k+1,j+1} + y_{i,k+1,j})$$

$$- R_i[\varphi_i(Y_{1,k+1,j+1} Y_{2,k+1,j+1}) + \varphi_i(Y_{1,k+1,j} Y_{2,k+1,j})],$$

$$R_i = \frac{x\varphi_i l}{2}; \qquad \frac{\partial y_i}{\partial \xi} = -\psi_i[y_i - \varphi_i(Y_1, Y_2)]$$

$$\frac{y_{i,k+1,j+1} - y_{i,k,j+1}}{2} = \psi_i \left\{\frac{y_{i,k+1,j+1} + y_{i,k,j+1}}{2}\right.$$

$$\left. - \frac{1}{2} [\varphi_i(Y_{1,k+1,j+1} Y_{2,k+1,j+1}) + \varphi_i(Y_{1,k,j+1}, Y_{2,k,j+1})]\right\}$$

$$y_{i,k+1,j+1} = \Theta_i y_{i,k,j+1} + P_i[\varphi_i(Y_{1,k+1,j+1} Y_{2,k+1,j+1}) + \varphi_i(Y_{1,k,j+1} Y_{2,k,j+1})]$$

$$\Theta_i = \frac{2 - \psi_i h}{2 + \psi_i h}; \quad P_i = \frac{\psi_i h}{2 + \psi_i h}.$$

Then in order to determine the initial functions Y_i and y_i at the next point it is possible to apply numerical equations

$$y_{i,k+1,j+1} = \Theta_i y_{i,k,j+1} + P_i[\varphi_i(Y_{1,k+1,j+1} Y_{2,k+1,j+1}) + \varphi_i(Y_{1,k+1,j+1} y_{2,k,j+1})] \tag{41}$$

$$Y_{i,k+1,j+1} = Y_{i,k+1,j} + R_i(y_{i,k+1,j+1} y_{i,k+1,j}) - R_i[\varphi_i(Y_{1,k+1,j+1} Y_{2,k+1,j+1}) - \varphi_i(Y_{1,k+1,j} Y_{2,k+1,j})] \tag{42}$$

The calculation begins from $k + 1, j + 1$ initial approximation $y_{i,k+1,j+1}$ determination by relationship (41) in so doing $\varphi_i(Y_{i,k+1,j}\ Y_{2,k+1,j})$ is supposed to be equal to $\varphi_i(Y_{1,k,j+1}\ Y_{2,k,j+1})$. $Y_{i,k+1,j+1}$ value is estimated under $\varphi_i(Y_{1,k+1,j+1}) = \varphi_i(Y_{i,k+1,j})$ by expression (42).

Then second refinements of values Y_i^k and y_i^k at points $k + 1 \cdot j + 1$ are determined. The number of refinements is selected by employing numerical experiment and in the given problem can be assumed to be equal to 3. Values Y_i and y_i along the coordinate axes required for calculation are taken from initial and boundary conditions (34)—(35). Determination of $Y_{i,0,j+1}$ is accomplished by the integration of system (36) by the Runge—Kutta method.

After the calculation of Y_i and y_i having been accomplished in the next time layer $j + 1$ a comparison of the relative concentrations of the poorly sorbed component at the output of the adsorbent $y_{1,k,j+1}$ is carried out. As a result of the calculations concentration distribution of all the components in solid and flowing phases along the length of the layer and their breakthrough curves behind the adsorbent layer are determined. In dividing a considerable number of components the necessity of kinetic constant estimating in experiments especially carried out complicates considerably non-equilibrium model application.

As multicomponent mixture separation can be accomplished effectively in the adsorbent layer of large length in which the process approaches equilibrium one estimation problem is simplified considerably if the adsorption process is assumed to be in equilibrium for all the components and process non-equilibrium is considered only for the poorly sorbing component. Let us consider briefly the formulation of the problem.

A mixture of various components with initial concentrations $C_{i0} (i = 1, 2 \ldots n)$ is supplied into the column of indefinitely large length. The components have various sorptivities and form a series of 1, 2 ... n. At the input of the column the nth substance with the highest sorptivity forms a saturated zone with a direct slope. In the same n region all other 1, 2, ... $(n - 1)$ substances are sorbed.

The $(n - 1)$ region will contain the $(n - 1)$ component only. As the nth component is not in this region, any longer substance distribution in it will be different. The front boundary speed of every zone is defined by that of the mostsorbing substance front. Having introduced the dimensionless parameters and variables

$$y_{i,j} = \frac{c_{ij}}{C_i n}; \quad Y_{i,j} = \frac{a_{ij}}{a_i^0}; \quad \xi = \frac{x}{L_0}; \quad \tau = \frac{Ut}{L};$$

$$P_i = \frac{A_i}{\Gamma_i c_{in}}; \quad \bar{v} = \frac{v}{u}; \quad g_i = \frac{a_i}{c_{in}}.$$

We obtained the mixture adsorption isotherm equation in the following form

$$Y_{i,n} = \frac{(1 + p_i)\, y_i/p_i}{1 + \sum_1^n y_i/p_i} \tag{43}$$

The dimensionless front speed and region length will be

$$\bar{v}_{i,i} = \frac{y_{i,i}}{y_{i,i} + g_{i,i}} \tag{44}$$

$$\Delta\xi = \xi_i - \xi_{i+1} = (\bar{v}_i - \bar{v}_{i+1})\,\tau \tag{45}$$

The compound concentration in the region will be defined by the relationship

$$y_{k,i} = y_{k,j+1} + \frac{g_k y_{i+1,i+1}}{g_{i+1} Y_{i+1,i+n}} \cdot (Y_{k,i} - Y_{k,i+1}) \tag{46}$$

The component concentration estimation in gaseous and solid phases of ith region is reduced to the solution of the non-linear equation system of balance (46) and adsorption isotherm (43). In order to solve the system let us use the method of the steepest descent. Let us consider the quadratic function

$$\varphi(y) = \sum_{k=1}^{i} G_k^2$$

where

$$G_k = y_{k,i+1} + \frac{g_k y_{i+1,i+1}}{g_{i+1} Y_{i+1,i+1}} (Y_{k,i} - Y_{k,i+1}), \tag{47}$$

acquiring minimum values of argument y, which satisfies all the equations of the system (47). Defining initial approximations y_k in i^{th} region ($y_{k,i} = 1, 2, y_{k,i+1,j}$) let us find roots of the system (minimum φ) by iteration method determining the rth approximation of the root by the relationship

$$y_{k,i}^r = y_{k,i}^{r-1} - \frac{2\varphi(y^{r-1})}{\sum_{k=1}^{i}\left[\frac{\partial}{\partial y_{k,i}}\varphi(y^{r-1})\right]^2} \frac{\partial\varphi(y^{r-1})}{\partial y_{k,i}} \tag{48}$$

By the differentiation of expression (47) and taking into consideration equation (46) we obtain the relationship for the estimation of

$$\frac{\partial \varphi(y)}{\partial y_{k,i}} = 2\left[-G_k + \sum_1^i G \frac{y_i y_{i+1,j+1}}{y_{i+1} Y_{i+1,j+1}} \frac{\partial Y_{j,i}}{\partial y_{k,i}}\right] \tag{49}$$

$$\frac{\partial Y_{l,i,k}}{\partial y_k} = -\frac{(1+p_i)\, y_l}{P_l p_k \left[1 + \sum_1^i y_i/p_i\right]^2} \quad \text{when } L \neq k$$

$$\frac{\partial Y_{l,i,k}}{\partial y_k} = \frac{(1+p_k)\left(1 - y_k/p_k + \sum_1^i y_i/p_i\right)}{p_k\left(1 + \sum_1^i y_i/p_i\right)^2} \quad \text{when } L = k$$

After every refinement of y_i the deviation of G from 0 is estimated. As soon as $\frac{1}{i} \sum |G_j| < \varepsilon$ calculation is ceased and the second root approximation is taken for the solution of the system (48). Overall calculation is ceased when definite breakthrough concentration of the poorly adsorbed component appears in the effluent. As the process of less sorbing component adsorption is assumed to be non-equilibrium the consideration of non-equilibrium can be accomplished by employing the equation of GUCHOVITSKY, ZABEGINSKY and TICHONOV [10].

$$\Theta = \frac{a_0'}{u c_0'}\left\{L - \frac{U}{\beta}\left[\frac{A}{\Gamma c_0'} \ln\left(\frac{c_0'}{c} - 1\right) + \ln\left(\frac{c_0'}{c}\right) - 1\right]\right\} \tag{50}$$

where:

β — mass-exchange coefficient of the less sorbing component;
c_0' — concentration of the less sorbing component arriving at zone 2 from zone 1;
c — defined breakthrough concentration.

By defining various values of less sorbing component breakthrough concentrations it is possible to calculate its output curves for various layer lengths and flowing phase rate. As a result of the calculation the speed and width of every zone, the values of adsorption and concentrations of all the mixture components in the zones, and the time of the sorbent activity are determined for the less sorbed component up to the defined breakthrough concentrations. Let us consider some results obtained in investigating two-component system adsorption by employing a digital computer. Figure 3 shows estimation results of output curves under various values of mass-transfer coefficients.

The change of range of both values β_1 and β_2 and β_1/β_2 relationship covers the possible change interval of these coefficients found in practice. When $\beta_1 = \beta_2 = 5\ \text{s}^{-1}$ the breakthrough curves of both components are rather extended and the breakthrough occurs roughly at the same time. Due to low mass-transfer coefficient values the breakthrough curve of the poorly adsorbed component is rather extended and the mass-transfer zone length moves over the apparatus. When the mass-transfer coefficient of the preferentially adsorbed component ($\beta_1 = 20$, $\beta_2 = 5\ s^{-1}$) increases intense contraction of the first component front is observed, which in its turn affects its breakthrough time and somewhat enlarges C_{2max}. The further growth of β_1 still increases the first component breakthrough time and practically does not influence the breakthrough curve of the second one. Thus, when β_2 values are small β_1 influence on the second component output curve is insignificant. The second component mass transfer coefficient increasing from 10 to 30 s^{-1} results in significant breakthrough time growth. The change from 10 to 30 s^{-1} leads to the total contraction of the second component front and to maximum concentration growth. Thus when the values of β_2 are low the influence of the values of β_1

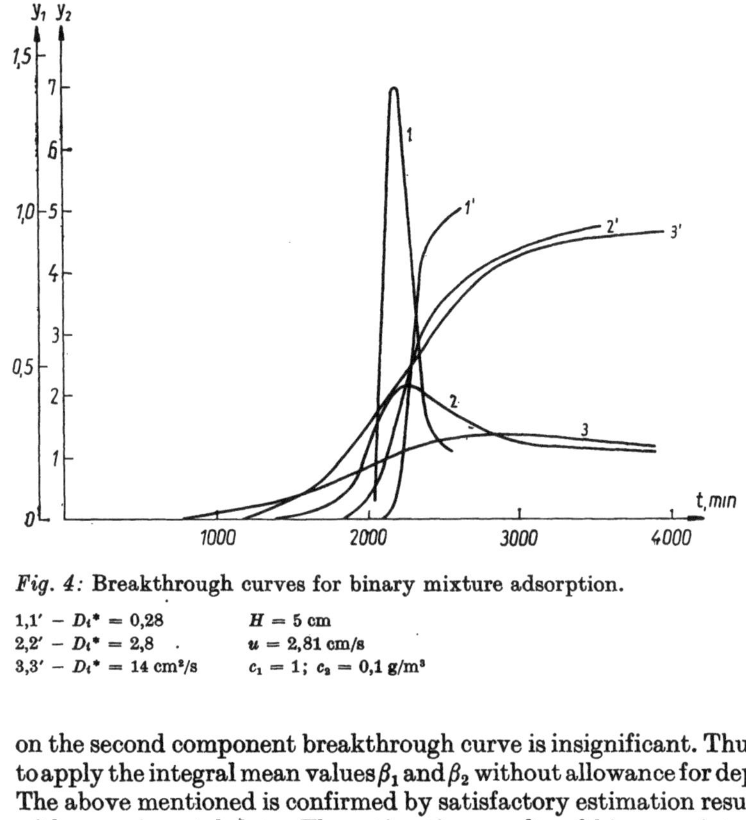

Fig. 4: Breakthrough curves for binary mixture adsorption.

1,1′ – $D_t^* = 0{,}28$	$H = 5$ cm
2,2′ – $D_t^* = 2{,}8$	$u = 2{,}81$ cm/s
3,3′ – $D_t^* = 14\ cm^2/s$	$c_1 = 1$; $c_2 = 0{,}1\ g/m^3$

on the second component breakthrough curve is insignificant. Thus it is possible to apply the integral mean values β_1 and β_2 without allowance for dependence $\beta(a)$. The above mentioned is confirmed by satisfactory estimation result coinoidence with experimental data. The estimation results of binary mixture for various values of D^* are shown on Fig. 4. In the successive diminishing of the values of D^* from 14 to 0.28 cm^2/s a marked smaller concentration of the sorbed component and a considerable contraction of the more sorbed component in the massexchange zone are observed.

Fig. 5 demonstrates experimental and calculated concentration distribution curves of the mixture of amber, glutaric, adipinic, azelaic acids in solid phase.

For the layers of such a length the mixture adsorption process is close to the equilibrium one. This is confirmed by satisfactory coincidence of estimated and experimental data. Applying mathematical models is particularly effective in investigating separation processes in adsorbent layers of large length proceeding for a long time. The estimation of such processes by a digital computer with the help of a combined model takes several seconds.

In carrying out displacing desorption processes as distinguished from component mixture absorption the desorbing agent is supplied with definite speed and concentration into the column uniformly filled with adsorbing substance.

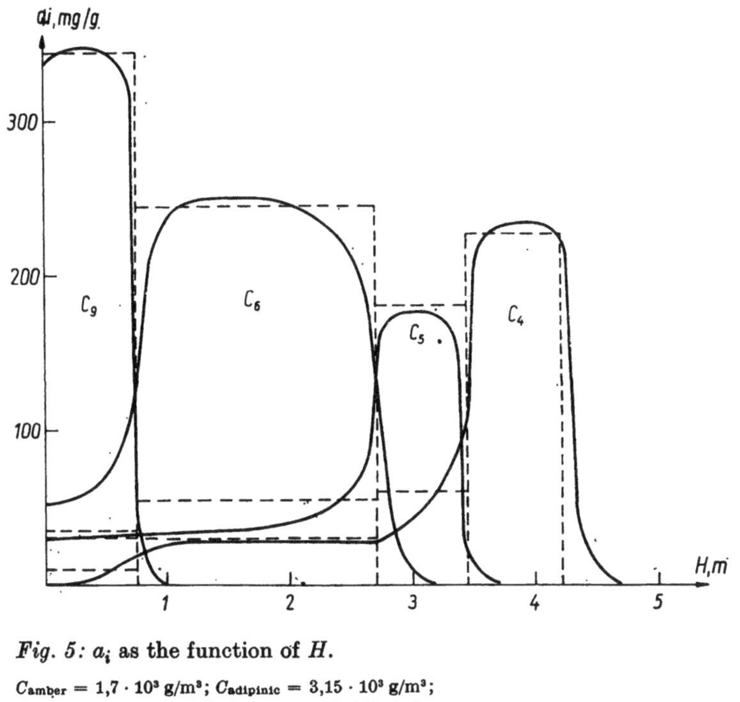

Fig. 5: a_i as the function of H.

$C_{amber} = 1{,}7 \cdot 10^3$ g/m^3; $C_{adipinic} = 3{,}15 \cdot 10^3$ g/m^3;
$C_{glutaric} = 1{,}5 \cdot 10^3$ g/m^3; $C_{azelaic} = 1{,}5 \cdot 10^3$ g/m^3;
$H = 4{,}7$ m, $u = 0{,}225$ cm/s;
– – – theory

In so doing the problem of calculation of concentration distribution and adsorption values along the layer length at any moment of time arose. In case of applying substances with high sorptivity as desorbing agents not only displacement but also concentrating occur and a regime providing the highest concentration has proved to be the object of design.

In using substances as displacing agents contaminating the evolved product the necessity of layer height determination providing the highest concentration without the breakthrough of the displacing agent arises. In order to investigate displacing processes mixture adsorption mathematical models can be applied. In employing the equilibrium model for the displacing process the boundary conditions are:

$$t = 0; \quad K \geqq x \geqq 0; \; c_2 = c_{02}; \; a_2 = a_{02}; \; c_1 = a_1 = 0. \tag{51}$$

$$t > 0 \begin{cases} x = 0; \; uc_{01} = uc_1(0) - D_i^* \dfrac{dc_i}{dx}; \; uc_1(0) = D_2^* \dfrac{dc_2}{dx} \\ x = H; \; D_2^* \cdot \dfrac{dc_i}{dx} = 0 \end{cases} \tag{52}$$

and for the non-equilibrium model

$$t = 0 \quad c_1 = a_1 = 0; \quad c_2 = c_{02}; \quad a_2 = a_{02} \tag{53}$$

$$t > 0 \quad x = 0 \qquad c_1(0, t) = 1; \quad c_2(0, t) = 0 \tag{54}$$

Fig. 6 shows designed output adsorption curves of water vapours and benzene desorption on the zeolite NaX. In the same figure experimental output curves are represented for comparison. Correspondence of the experiment and design is satisfactory. In order to determine the displacing process dynamics peculiarities by employing the digital computer the investigation of adsorbing

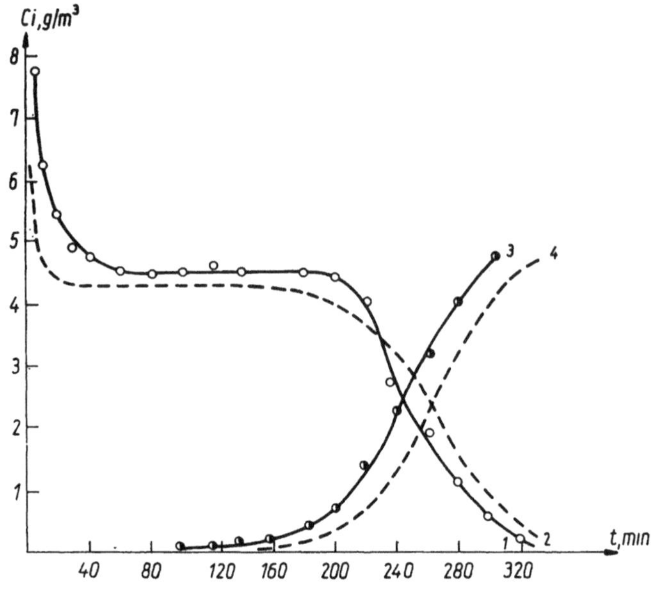

Fig. 6: Benzene desorption from NaX by water vapour.

– – theory
—— experiment
1,2 – C_6H_6; *3,4* – H_2O;
$H = 25$ cm, $u = 28{,}1$ cm/s, $C_{H_2O} = 5$ g/m³

properties of displacing agents and adsorbents was carried out. Fig. 7 shows curves of n-octane displaced from Zeolite NaX by water vapours under various concentrations of the displacing agent. When the water vapour concentration is 5 g/m^3 the time necessary for the total evolving of n-octane is 200 min, the n-octane concentration behind the zeolite layer being increased to 2.8 g/m^3, when the water vapour concentration is 11 g/m^3 and 30 g/m^3 the desorption

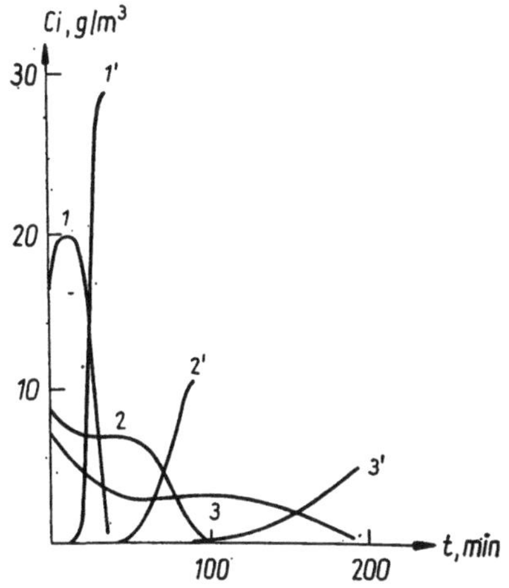

Fig. 7: n-octane desorption from NaX by water vapours.
C_{H_2O}; *1* – 30 g/m^3, *2* – 11 g/m^3, *3* – 5 g/m^3;
H = 15 cm; u = 28,1 cm/s

process comes to an end at 90 and 35 min, respectively. The results of such estimations can be applied for the evaluation of the time necessary for the desorption of hydrocarbons and concentration extent determination. Desorption extent estimation results under various values of Henry's coefficient are represented in Tab. 2. It follows from these data that when sorptivity decreases hundred fold the blowing off of the investigated component from the layer rather than displacing is observed, the rate of this process being high only at early stages when the removal of less bound molecules occurs. The desorption

Table 2: The influence of Γ on the amount desorbed .

Time, min	Amount desorbed, %		
	$\Gamma = 0.65 \cdot 10^6$	$\Gamma = 0.65 \cdot 10^5$	$\Gamma = 0.65 \cdot 10^4$
55	80.5	62.5	41.4
85	99,5	76.5	49.6
95	100	79.8	52.0

of the rest molecules proceeds with negligible rate. As at present methods of mass transfer coefficient evaluation of single component adsorbed on the adsorbent filled with an other component are missing individual substance mass exchange coefficients were used in estimations of displacing processes [8]. Fig. 8 displays the results of n-octane displacement by water vapour. In the same figure displacement curves estimated for the same parameters a and g but for different mass transfer coefficients are shown. The change range of β_1 and β_2 and relationships β_1/β_2 as well cover possible change intervals of these coefficients. When $\beta_1 = 15$ the curves differ from each other insignificantly (there is a difference only in the low breakthrough concentration region).

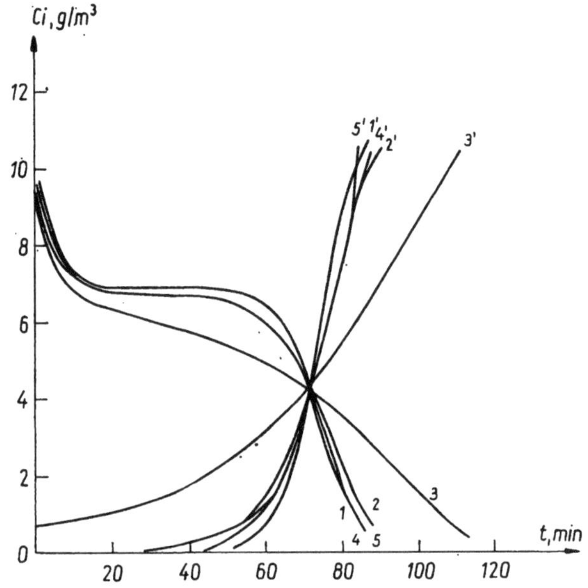

Fig. 8: Breakthrough curves.

1,1' – $\beta_1 = 30$, $\beta_2 = 5$; *2,2'* – $\beta_1 = 18$, $\beta_2 = 5$;
3,3' – $\beta_1 = 5$, $\beta_2 = 5$; *4,4'* – $\beta_1 = 15$, $\beta_2 = 15$;
5,5' – $\beta_1 = 15$, $\beta_2 = 30$ 1/s
$H = 15$ cm, $C_1 = 11$ g/m³, $u = 28{,}1$ cm/s

The output curves of the displaced component are similar to the ones mentioned above. When mass transfer coefficients are equal but rather high ($\beta_1 = \beta_2 = 15$) the displaced substance desorption curves practically coincide with the curves for which $\beta_1/\beta_2 = 6$ (curve *1*) and $\beta_1/\beta_2 = 1/2$ (curve *5*). In case when kinetic properties of the displacing agent are insignificant (curve *3* $\beta_1 = \beta_2 = 5$) rapid breakthrough of the displacing agent and considerable curve distortion of both components are observed. Thus it follows from the obtained data that the shape of the curve of the displaced compound is influenced primarily by the mass transfer coefficient of the displacing agent; the contribution of the latter to the displacement dynamics is insignificant. Taking into consideration the slight influence of β_2 upon the output curves of both components it is possible to use in the first approximation for such estimations mass transfer coefficient average integral values evaluated by employing output curves of the individual substances, which is confirmed by the coincidence of estimation results and experimental data.

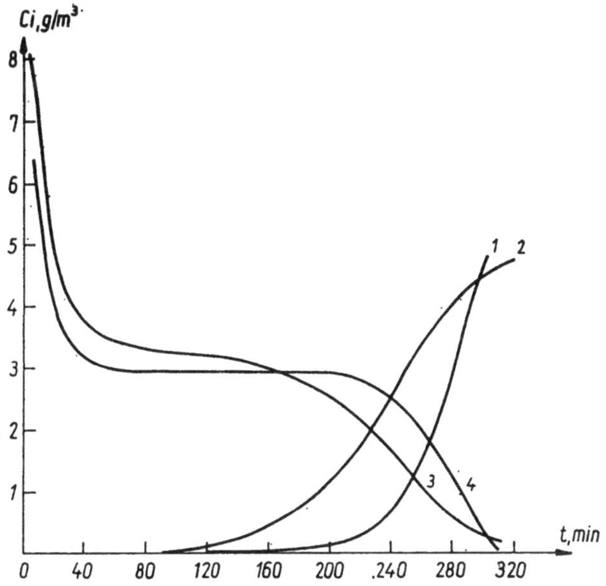

Fig. 9: n-octane desorption from NaX by benzene and water.

1 – H_2O; *2* – C_6H_6; *3,4* – C_8H_{18}
$u = 28{,}1$ cm/s, $H = 25$ cm, $C_1 = 5$ g/m³

Fig. 9 shows results obtained in simulating component displacement by agents differing in sorptivity and kinetic characteristics. In applying benzene vapour as a displacing agent its breakthrough curves are more expanded and accordingly the n-octane breakthrough curves are more extended. In using water vapour the n-octane breakthrough curves contract.

References

[1] RACHINSKI, V. V., "*Introduction into General Theory of Adsorption Dynamics and Chromatography*", Nauka, Moscow 1964.
[2] TODES, O. M., "*Kinetics and Dynamics of Physical Adsorption*". Nauka, Moscow 1973.
[3] "*Calculation and Optimization Fundamentals for Ion-exchange Processes*", Nauka, Moscow 1973

[4] CLASSON, C., "*Adsorption Analysis of Mixtures*", Goschimizdat, Moscow—Leningrad 1950.
[5] SEBALLO, A. A., PLACHENOV, T. G., LESIN, YU. S., BARANOV, E. I., "*Kinetics and Dynamics of Physical Adsorption*", 275, Nauka, Moskow 1973
[6] MYERS, A. L., PRAUSNITZ, J. M., Amer. J. Eng. I. **11**, 121, (1965).
[7] BERING, B. P., SERPINSKI, V. V., SURINOVA, S. I., DAN USSR **153**, 1, 129 (1963).
[8] PLACHENOV, T. G., SEBALLO, A. A., BARANOV, E. I., SHIRYAEV, A. N., J. Appl. Chem. (USSR) **43**, 4, 816, (1970).
[9] LEVENSPIEL, O., Chemical Reaction Engineering, Wiley, New York, London, 1965.
[10] JUCHOVITSKI, A. A., ZABEJINSKI, YU. L., TICHONOV, A. N., J. Phys. Chem. (USSR) **19**, 253, (1945).

On mass-exchange influence upon mass-transfer in fixed granular bed

(E. A. Ustinov, V. L. Kolin, I. N. Taganov)

Zusammenfassung:

Es wird ein Modell des Gasstromes untersucht, in dem einzelne Teile turbulenten Fluktuationen unterliegen, wobei die elementare Verschiebung in bzw. gegen die Strömungsrichtung verschiedene Wahrscheinlichkeit besitzt.

Die für dieses Modell abgeleitete Differentialgleichung für den Transport und die Randbedingungen berücksichtigt die Konzentrationsverteilung des Sorptivs in der festen Phase. Es wird eine Methode zur näherungsweisen Berücksichtigung von Längsvermischungseffekten bei der Modellierung von Adsorptionsprozessen in Form eines Systems von Differentialgleichungen hyperbolischen Typs angeführt. Außerdem werden Lösungsbeispiele und ein Vergleich mit anderen Modellen angegeben.

1. Introduction

In order to describe sorbed substance transfer in a flow through the granular bed Einstein-Kolmogorov's differential equation is generally used

$$(1-\varepsilon)\frac{\partial a}{\partial t}+\varepsilon\frac{\partial c}{\partial t}+v\frac{\partial c}{\partial x}=D_E\frac{\partial^2 c}{\partial x^2} \tag{1}$$

It connects the longitudinal dispersion effect mainly with hydrodynamic conditions, in particular under non-uniformity in grain packing and turbulence, [1—5]. As in the transfer in homogeneous media value D_E can be expressed with the aid of turbulence scale λ and relaxation time t_{rel}. If longitudinal dispersion is connected mainly with speed turbulent fluctuations of single elements of the flow then

$$D_E \cong \frac{1}{3}\,\varepsilon\,\frac{\lambda^2}{t_{rel}}. \tag{2}$$

It is assumed here that λ and t_{rel} are equal to the length and time of continuous movement in the defined direction both of flow particles and sorbing substance molecules. However, in effect these values of flow and sorbing substance are significantly different due to mass-exchange, and its longitudinal transfer is likely to depend on mass exchange rate. On the other hand, longitudinal dispersion is contributed to by sorbed substance diffusion along the flow movement direction within individual sorbent grains, and the greater this contribution the more intense the mass exchange. Known theoretical concepts of longitudinal transfer in the flow through the layer of sorbing granular material [1, 2] ignore this fact and assume that the flow structure does not change in passing from non-sorbing tracer to sorbing component if hydrodynamic conditions of flow movement are constant. Mass-exchange influence upon the longitudinal transfer of the component to be sorbed in a sorbing practicle bed of solid material is the subject of the present paper.

2. Physical model

Let us consider the isothermic process of substance absorption from a large volume of incompressible inert gas-carrier moving with constant average speed through a motionless bed of sorbing granular material. Mass-exchange and mass-transfer are accomplished by employing individual flow elements which due to turbulence undergo fluctuations with constant speed value relative to the bed, but possess various probabilities of movement. Average length λ and time t_{rel} of the movement of the elements are of the same values in all directions. Dimensions of the elements are similar to the turbulence scale λ.

3. General equations

Let us isolate a bed element of unit cross section and thickness dx. The surface area of the grains distributed over the space from $x - R$ to $x + R$ in the isolated volume is $3(1 - \varepsilon)\, dx/dR$. The number of elements with this surface at the unity of time is equal to $v_{fl}(1 - \varepsilon)\, dx/\lambda^3 d$. The time of continuous contact of the elements with the surface similar to relaxation time t_{rel} is assumed to be sufficiently short, so that the relative concentration change of the adsorbed substance inside the elements in this period of time will be insiginficant as well. In this case mass-exchange rate of an element and sorbent is

$$q_m = -\lambda D_e \frac{\partial C}{\partial n}\bigg|_s = \lambda^2 \sqrt{\frac{D_e}{\pi t}}\,(C - \bar{C}_s^*) \tag{3}$$

where n — normal to the mass-exchange area;

$\bar{c}_s^* = \frac{1}{d} \int\limits_{x-R}^{x+R} c_s^*(x_1, x)\, dx_1,$ $c_s^*(x_1, x)$ — is the concentration of the sorbing substance in the flow equilibrium with sorption value a_s at point x on the grain surface, the centre of which is situated at point x_1.

The quantity of the substance passed through the contact surface of the sorbent and the element over the period of time t_{rel} is

$$\int\limits_0^{t_{rel}} q_m dt = 2\lambda^2 \sqrt{\frac{D_e t_{rel}}{\pi}}\,(C - \bar{C}_s^*), \tag{4}$$

mass exchange rate in the isolated bed element being

$$Q_m dx = (1 - \varepsilon)\, \beta_e dx (C - \bar{C}_s^*), \qquad \beta_e = \frac{2}{d} \sqrt{\frac{D_e v_{fl}}{\pi \lambda}}, \tag{5}$$

The internal diffusion equation of the individual grain is the following

$$\frac{\partial a_i}{\partial t} = D_i \frac{\partial^2 a_i}{\partial r^2} + D_i \frac{1}{r} \frac{\partial a_i}{\partial r} + D_i \frac{\partial^2 a_i}{\partial x^2}. \tag{6}$$

Then we obtain the equation for the average sorption for the grain bed section

$$\frac{\partial \bar{a}}{\partial t} = D_i \frac{\partial^2 \bar{a}}{\partial x^2} + \beta_e(C - \bar{C}_s*) \tag{7}$$

in this case

$$\left.\frac{\partial a_i}{\partial r}\right|_{r=0} = 0{,}3(R^2 - l^2)^{1/2} \frac{D_i}{R^2} \left.\frac{\partial a_i}{\partial r}\right|_{r=\sqrt{R^2-l^2}} = \beta_e(C - C_s*),\ l = x - x_1 \tag{8}$$

Let us isolate from the overall flow q_x the two constituents q_{1x} and q_{2x}, in which molecules move only in direct or opposite direction, respectively. Any molecule in every flow not only changes the direction of movement at the cost of fluctuations but is adsorbed with the greater mass exchange rate. This results in the decreasing of average length and time values of continuous molecule movement in the defined direction in comparison with λ and t_{rel}. On the other hand, there are not only the particles mentioned above but also the particles desorbed from the immobile phase which continuously enter flows q_{1x} and q_{2x}. Therefore mass exchange should result in the distortion of differential equations describing the behaviour of flows q_{1x} and q_{2x}.

The probability of particle movement direction change over the unit of bed length is inversely proportional to turbulence scale. The probability of adsorption is proportional to substance adsorption rate in the isolated bed volume without allowance of desorption. Then the relative decrease of the flows q_{1x} and q_{2x} over the area of unit length is $\frac{1}{\lambda} + \frac{1-\varepsilon}{\varepsilon} \beta_e \frac{t_{rel}}{\lambda}$. On the other hand, in the same area of the bed flows q_{1x} and q_{2x} are increased at the cost of the molecules entered into them from both phases. Every molecule changes, on an average, its direction per unit time 1/trel times, the desorption rate at the defined bed element being equal to $\frac{1-\varepsilon}{\varepsilon} \beta_e \bar{C}_s*$. Taking into consideration that the probabilities of the subsequent molecule movement of increasing and decreasing of x are equal to p_1 and p_2 respectively, we obtain the following differential equations for flows q_{1x} and q_{2x}:

$$\lambda \frac{\partial q_{1x}}{\partial x} + t_{rel} \frac{\partial q_{1x}}{\partial t} + \left(1 + \frac{1-\varepsilon}{\varepsilon} \beta_e t_{rel}\right) q_{1x} = p_1 v_{fl} C + p_1 \frac{1-\varepsilon}{\varepsilon} \beta_e \lambda \bar{C}_s* \tag{9}$$

$$-\lambda \frac{\partial q_{2x}}{\partial x} + t_{rel} \frac{\partial q_{2x}}{\partial t} + \left(1 + \frac{1-\varepsilon}{\varepsilon} \beta_e t_{rel}\right) q_{2x} = p_2 v_{fl} C + p_2 \frac{1-\varepsilon}{\varepsilon} \beta_e \lambda \bar{C}_s* \tag{10}$$

Evidently the material balance equation is:

$$\varepsilon \frac{\partial q_x}{\partial x} + \varepsilon \frac{\partial C}{\partial t} + Q_m = \varepsilon D_e \frac{\partial^2 C}{\partial x^2}. \tag{11}$$

In this case contribution to molecule diffusion longitudinal transfer in the direction of the flow movement inside the elements is taken into consideration. As the movement speed of individual molecules is rather high in comparison with the turbulence scale, the mass exchange practically does not influence the molecular consistuent of longitudinal transfer.

System (9)—(11) is considerably simplified if the distribution coefficient $\Gamma = a_0/c_0$ is sufficiently high, which is common to sorption processes. In this case derivates in respect of time in (9)—(11) can be neglected and after simple transformations it follows from (9)—(11) that

$$v\frac{\partial C}{\partial x} + v(1-\alpha)\frac{\partial(C-\overline{C}_s^*)}{\partial x} + \varepsilon\frac{1-\alpha}{\alpha}\frac{v_{fl}}{\lambda}(C-\overline{C}_s^*)$$
$$= (\alpha D_E + \varepsilon D_e)\frac{\partial^2 C}{\partial x^2} + 2\alpha(1-\alpha)\,D_E\frac{\partial^2(C-\overline{C}_s^*)}{\partial x^2} \tag{12}$$

where:

$$\alpha = \frac{\lambda_1}{\lambda} = \frac{\varepsilon v_{fl}}{\varepsilon v_{fl} + (1-\varepsilon)\,\beta_e\lambda};$$

λ_1 — average length of sorbate molecule free dislocation;

$D_E = \frac{1}{3}\,\varepsilon\,\frac{\lambda^2}{t_{rel}}$ — longitudinal transfer coefficient when mass-exchange is missing.

Considering (7) instead of (12) gives:

$$v\frac{\partial C}{\partial x} + (1-\varepsilon)\frac{\partial\bar{a}}{\partial t} = v(1-\alpha)\frac{\partial(C-\overline{C}_s^*)}{\partial x} + (\alpha D_E + \varepsilon D_e)\frac{\partial^2 C}{\partial x^2}$$
$$+ 2\alpha(1-\alpha)\,D_E\frac{\partial^2(C-\overline{C}_s^*)}{\partial x^2} + (1-\varepsilon)\,D_i\frac{\partial^2\bar{a}}{\partial x^2}. \tag{13}$$

Thus instead of transfer (1) and kinetics (2) equations we obtain the system of equation (6), (7), (13). The terms in the right part of equation (13) give rise to quasidiffusion effects in mass-exchange in granular bed.

When in (13), $\lambda \to 0$, $\alpha \to 1$, $D_E \to 0$ we obtain

$$v\frac{\partial C}{\partial x} + (1-\varepsilon)\frac{\partial\bar{a}}{\partial t} = (1-\varepsilon)\,D_i\frac{\partial^2\bar{a}}{\partial x^2}. \tag{14}$$

Consequently in this case longitudinal effects also take place and stem from of the mass-exchange fact itself, which is possible only when $D_i \neq 0$, the influence of the longitudinal effect upon the mass-transfer becaming more significant with intradiffusional resistance to mass-exchange.

If the mass-exchange rate is limited by the internal diffusion rate then $C_s^* = C$ and

$$v\frac{\partial C}{\partial x} + (1-\varepsilon)\frac{\partial\bar{a}}{\partial t} = (1-\varepsilon)\,D_i\frac{\partial^2\bar{a}}{\partial x^2} + (\alpha D_E + \varepsilon D_e)\frac{\partial^2 C}{\partial x^2}. \tag{15}$$

If in so doing $D_e \to 0$, $\beta_e \to \infty$ then $\alpha \to 0$ and we return to (14).

Having substituted $C - \overline{C}_s{}^*$ differences in the right part (13) by their expression in equation (7) and neglected the derivatives of the order higher than the second instead of (13) we obtain approximately

$$v \frac{\partial C}{\partial x} + (1 - \varepsilon) \frac{\partial \bar{a}}{\partial x}$$
$$= \frac{1}{3} \frac{v\lambda^2}{D_E} \alpha(1 - \varepsilon) \frac{\partial^2 \bar{a}}{\partial x\, \partial t} + (\alpha D_E + \varepsilon D_e) \frac{\partial^2 C}{\partial x^2} + (1 - \varepsilon) D_i \frac{\partial^2 \bar{a}}{\partial x^2}. \tag{16}$$

We define the boundary conditions of (16) assuming that the concentration is discontinuous at the flow input in the bed and continuous at the bed output. Then we obtain from (9)—(11)

$$vC_{+0} = vC_0 + \frac{1}{3} \frac{v\lambda^2}{D_E} \alpha(1 - \varepsilon) \frac{\partial \bar{a}}{\partial t} + (\alpha D_E + \varepsilon D_e) \frac{\partial C}{\partial x} + (1 - \varepsilon) D_i \frac{\partial \bar{a}}{\partial x} \quad \text{when } x = 0 \tag{17}$$
$$\frac{1}{3} \frac{v\lambda^2}{D_e} \alpha(1 - \varepsilon) \frac{\partial \bar{a}}{\partial t} + (\alpha D_E + \varepsilon D_e) \frac{\partial C}{\partial x} + (1 - \varepsilon) D_i \frac{\partial \bar{a}}{\partial x} = 0 \quad \text{when } x = L.$$

It follows from (17) that the type of boundary conditions for the mass-transfer equation depends on the mass-exchange rate as well.

In the equation of kinetics (7) value $\overline{C}_s{}^*$ can be calculated by the joint solving of (6), (7) with boundary conditions (8). In the particular case at the asymptotic process stage or when D_i is significantly great and sorption extent change for grain volume is not large dependence of a_i upon r can be approximated by polynominal of the second order. Then we obtain

$$\beta_e(C - \overline{C}_s{}^*) = \frac{8D_i}{R^2} (\bar{a}_s - \bar{a}). \tag{18}$$

As $\bar{a}_s$ and $\overline{C}_s{}^*$ are connected with the isotherm equation, having evaluated $\overline{C}_s{}^*$ from (17) and substituted it into (7) we can obtain the equation of kinetics in the form

$$\frac{\partial \bar{a}}{\partial t} = D_i \frac{\partial^2 \bar{a}}{\partial x^2} + \beta_E(C - C^*). \tag{19}$$

Here the efficient mass-exchange coefficient β_E in the general case depends on a and C and is evaluated from the following expression:

$$\beta_E = \frac{\beta_e \beta_i}{\beta_e + \beta_i} \cdot \frac{C - \dfrac{\overline{C}_s{}^*}{\bar{a}_s} a}{C - C^*}, \quad \beta_i = \frac{8D_i}{R^2} \frac{\bar{a}_s}{\overline{C}_s{}^*}. \tag{20}$$

When the change of the sorption value for the grain radius is not abrupt it is possible to represent a_s by the TAYLOR series and use only two members of the expansion. Then instead of (20) we obtain approximately:

$$\beta_E = \frac{\beta_e \beta_i}{\beta_e + \beta_i}; \quad \beta_i = \frac{8D_i}{R^2} \frac{d\bar{a}}{d\bar{C}^*}. \tag{21}$$

It follows from (21) that with sorption extent increase under the convex isotherm β_E and β_i decrease.

In order to accomplish approximate registration of longitudinal effects let us introduce the subsidiary function

$$C_1 = C - \frac{1}{3} \frac{v\lambda^2}{D_E} \alpha(1-\varepsilon) \frac{\partial \bar{a}}{\partial t} - \left(\frac{D_E \alpha}{v} + \frac{\varepsilon D_e}{v}\right) \frac{\partial C}{\partial x} - \frac{(1-\varepsilon)\, D_i}{v} \frac{\partial \bar{a}}{\partial x}. \tag{22}$$

Then relative to C_1 and $\bar{a}$ instead of (16) we obtain the material balance equation containing no members resulting in longitudinal dispersion

$$v \frac{\partial C_1}{\partial x} + (1-\varepsilon) \frac{\partial \bar{a}}{\partial t} = 0 \tag{23}$$

the boundary conditions for C_1 acquiring the following form:

$$x = 0,\ C_{1+0} = 0; \qquad x = L,\ C_{1L-0} = C_L. \tag{24}$$

Comparing (16), (19), (22) and neglecting values of the second order we find approximately

$$C_1 = C + (1-\varepsilon)\, \beta_E \left[\frac{\alpha D_E + \dfrac{\varepsilon}{1-\varepsilon} D_e}{v^2} + \frac{D_i}{vu} - \frac{1}{3} \frac{\lambda^2}{D_E} \alpha \right] (C - \bar{C}^*) \tag{25}$$

$$\frac{\partial \bar{a}}{\partial t} = \beta_E^*(C_1 - \bar{C}^*). \tag{26}$$

Where u — concentrating point speed of movement along the bed.

In the equation of kinetics (26) the reduced mass exchange coefficient it evaluated from the following equation:

$$\frac{1}{\beta_E^*} = \frac{\alpha}{\beta_e} + \frac{1}{\beta_i} + \frac{(1-\varepsilon)\, \alpha D_E + \varepsilon D_e}{v^2} + \frac{(1-\varepsilon)\, D_i}{vu}. \tag{27}$$

Thus the system of equations (6), (7), (16) is approximately reduced to the system of equations (23), (26) of hyperbolic type relative to a and subsidiary function c_1, which are the solutions of the mass exchange dynamics problem under ideal displacement in the flow and the missing of sorbate longitudinal diffusion along the solid phase. Actual concentration in flow C is connected with expression (25). Longitudinal effects are approximately considered in the reduced mass exchange coefficient β_E^* by expression (27). The effective mass exchange coefficient β_E is evaluated by employing (21) with allowance for the isotherm equation.

4. Discussion of results

In analysing the obtained results let us take into consideration that $v_{fl} \sim v$ and $\lambda \sim d$. The speed of movement of points of equal concentration u is approximately constant at the asymptotic stage of front adsorption and in chromatographic processes. In the regions of progressing sorption front expansion, for example in the front process described convex isotherm $u = v \frac{\partial C^*}{\partial a}$.

The greatest influence upon the dynamic process is produced by longitudinal effects with low speeds of the flow. When $v \to 0$ it follows from (25) that concentrations in the low phases are connected by isotherm equation β_E, $\beta_E^* \to 0$ and $\beta_E \sim v^{0,5}$ $\beta_E^* \sim v^2$. When v is great $C_1 \to C$, $\beta_E^* \to \beta_i$ and the sorption rate is limited by the internal diffusion rate. The reduced mass-exchange coefficient β_E^* increases everywhere along with gas-carrier flow speed increase or grain diameter decrease. For convenience in discussion let us rewrite equations (25), (27) in dimensionless form

$$\varrho_E^* = \alpha\varrho_e + \varrho_i + \frac{(1-\varepsilon)\,k_\lambda}{3\Delta p}\,\frac{1}{h} + \left[\varepsilon + \frac{(1-\varepsilon)\,xu_0}{u}\right]\frac{1}{h^2} \tag{28}$$

$$y_1 = y + \left(\frac{\varrho_E^*}{\varrho_E} - 1\right)(y - y^*). \tag{29}$$

Where:

$h = \frac{vd}{D_e}$ — parameter proportional to relationship of the turbulent diffusion coefficient to the molecular one;

ϱ_i, ϱ_e, ϱ_{Ee} — values proportional to internal, external and reduced resistance and $\varrho_E = \varrho_e + \varrho_i$.

Fig. 1 shows dependence of dimensionless values $\frac{\beta_E d^2}{D_e}$, $\frac{\beta_E^* d^2}{D_e}$ upon $\sqrt{h}$. The figure shows that the difference between β_E and β_E^* is greatest with low v. When $v \to \infty$, β_E and β_E^* tend to one and the same limit β_i; however the change of β_E^* occurs significantly slower.

This is connected with masse-exchange influence upon mass-transfer, which even in great h results in an essential difference of turbulence scale λ and free dislocation length of sorbed substance molecules λ_1.

Fig. 2 displays the dependence of β_E and β_E^* upon grain diameter d. It is seen from the figure that β_E and β_E^* decrease abruptly with the increase of d, which can be accounted for by increasing internal diffusion resistance to mass-exchange.

The dependence of equivalent theoretical plates $\frac{v}{\beta_E^*}$ upon flow rate is shown in Fig. 3. The curve posses minimum, its descending branch being connected both with the molecular diffusion influence of substance tobe sorbeb in gas-carrier flow and the diffusion influence inside the singular bed grains.

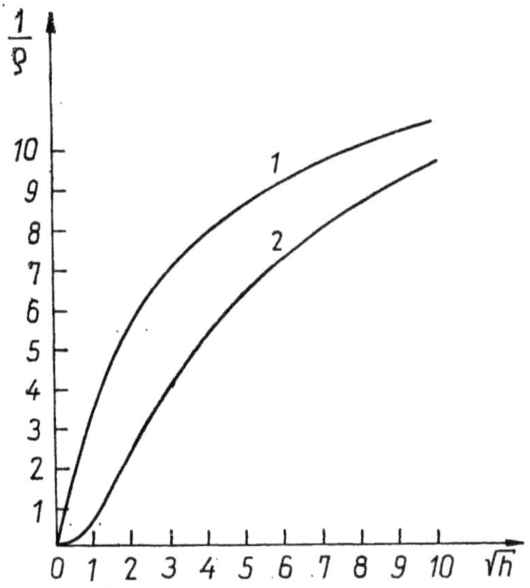

Fig. 1: Dependence of the value of the coefficient of mass-transfer on the rate of flow.

$E = 0{,}4$; $x = K_\lambda = 1$; $\Delta P = 1/6$; $P = \infty$, $1 - \beta_3(V)$ $2 - \beta_3{}^*(V)$

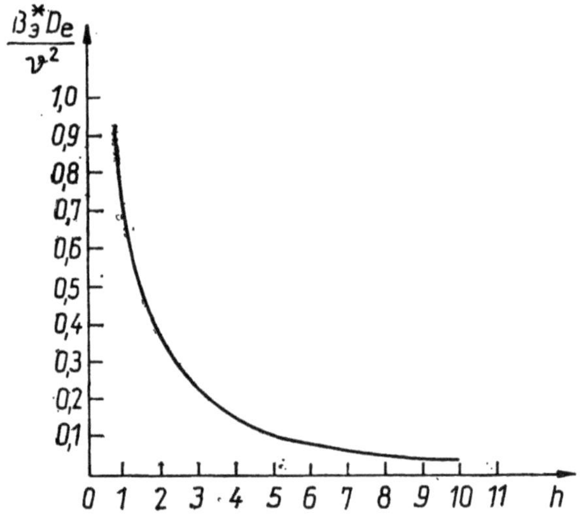

Fig. 2: Dependence of the value of effective coefficient of mass-transfer on the diameter of grains.

$\varepsilon = 0{,}4$; $x = K_\lambda = 1$; $\Delta P = 1/6$; $P = 0.5$; $\gamma = 0.5$.

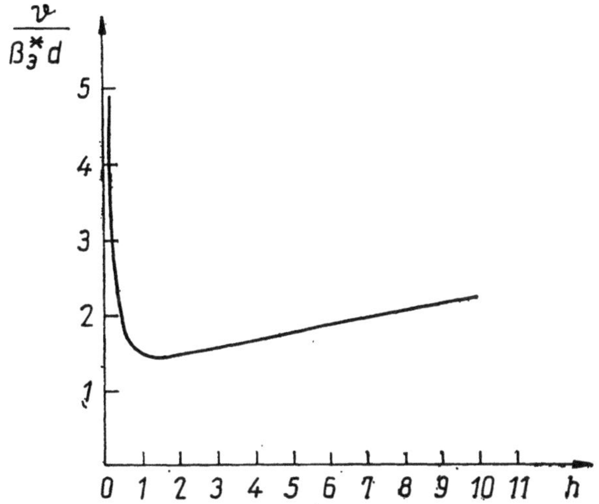

Fig. 3: Dependence of the effective theoretical plate on the rate of flow.
$\varepsilon = 0.4$; $x = K_\lambda = 1$; $\Delta P = 1/6$; $P = 0.5$; $\gamma = 0.5$.

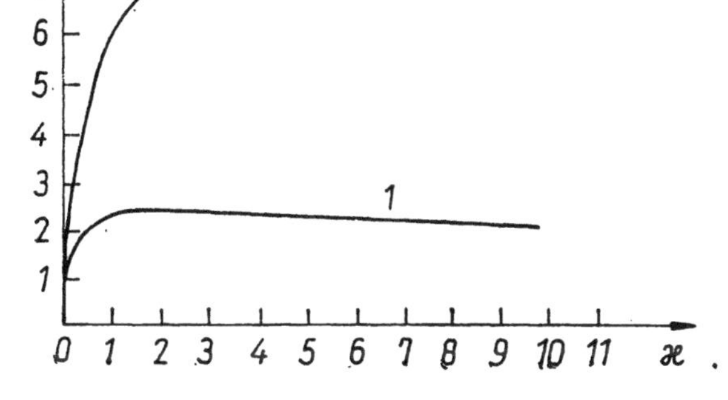

Fig. 4: Dependence of the coefficients of mass-transfer on the value of the diffusion coefficient.
$\varepsilon = 0.4$; $K_\lambda = 1$; $\Delta P = 1/6$; $P = 0.5$; $h = 4$; $\gamma = 0.5$;
1 — β_3^*(Di); 2 — β_3(Di).

When the internal diffusion coefficient increases, all other conditions being equal, resistance to internal diffusion mass-exchange naturally decreases, but β_i and β_E increase. However, longitudinal effects connected with sorbing substance diffusion along the bed inside the solid phase increase with D_i, which results in the fall of the reduced mass-exchange coefficient $\beta_E{}^*$ in comparison with β_E.

Fig. 4 shows that if $\beta_Э$ increases up to β_e with the increase of $\varkappa = \frac{D_i}{D_e} \times \left(\frac{d\bar{a}}{dC^*}\right)_{\bar{a}=0}$, maximum having been reached $\beta_E{}^*$ decreases tending asymptotically to 0. In so doing the maximum point $\beta_E{}^*$ corresponds to the following condition:

$$\varkappa_m{}^2(1-\varepsilon)\frac{u_0}{u} = h^2\frac{1}{32}\left(\frac{d\bar{a}}{d\bar{C}^*}\right)_{\bar{C}^*=0}\bigg/\frac{da}{dC},$$

When the sorption rate increases under convex isotherm and $D_i = \text{const}$ the internal mass-exchange coefficient β_i and the reference reduce, which results in $\beta_E{}^*$ diminution if u is constant. Dependences of $\frac{\beta_E{}^* d^2}{D_e}$ upon $y = \frac{c}{c_0}$ and $\gamma = \frac{a}{a_0}$ under various values of h for frontal adsorption with the Langmuir isotherm and $u = v\frac{c_0}{a_0}$ are in Fig. 5. It is seen from the figure that $\beta_E{}^*$

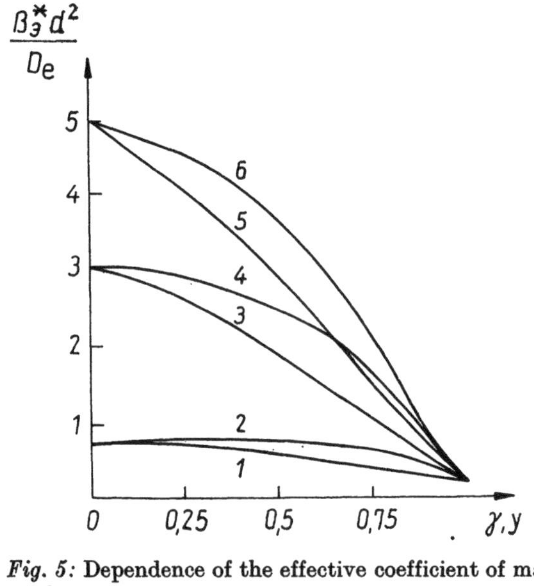

Fig. 5: Dependence of the effective coefficient of mass-transfer on the concentration

$\varepsilon = 0.4$; $x = K_\lambda = 1$; $\Delta P = 1/6$; $P = 0.1$; h. 1–1.2; 3.4–5; 5.6–10
1, 3, 5 – $\beta_з{}^*(y)$; 2, 4, 6 – $\beta_з{}^*(\gamma)$.

decreases along with γ and y, decreasing being the greater the higher h is. In case of low flow rate or small grain dimension d curves $\beta_E^* = \beta_E^*(\gamma)$ are convex and concave if grains are large; whereas in intermediate modification region v or d the curves are S — shaped.

In case of the progressive expansion of the sorption front, for example in frontal desorption the speed of concentration points increases with γ, which results in the diminution of the contribution of longitudinal effects connected with molecular transfer over the solid phase and the increase of the mass-exchange reduced coefficient β_E^*. In so doing curve $\beta_E^* = \beta_E^*(\gamma)$ reaches process minimum.

When D_i is varying the dependence of β_E^* upon γ can be of another form. Thus, if the internal dissufion rate is limited by the transfer of the adsorbed substance in macropores, internal diffusion coefficient D_i with γ growth increases rather abruptly, whereas β_i varies insignificantly when passing the minimum. At the same time the contribution of longitudinal effects from molecular longitudinal diffusion in the solid phase with the growth of γ increases and therefore β_E^* as in case of D_i = const decreases; however variations of β_E^*, and γ become less pronounced when the sorption kinetics passes to the intradiffusion region and flow v speed increases, when $v \to \infty$, β_E^* does not depend upon γ. On the contrary, when the v speed is low β_E^* diminution with γ growth should be more significant in spite of the fact that sorption in this case proceeds to the external diffusion region. Evidently this fact can result in the erroneous conclusion of the external mass-exchange coefficient β_e depend-

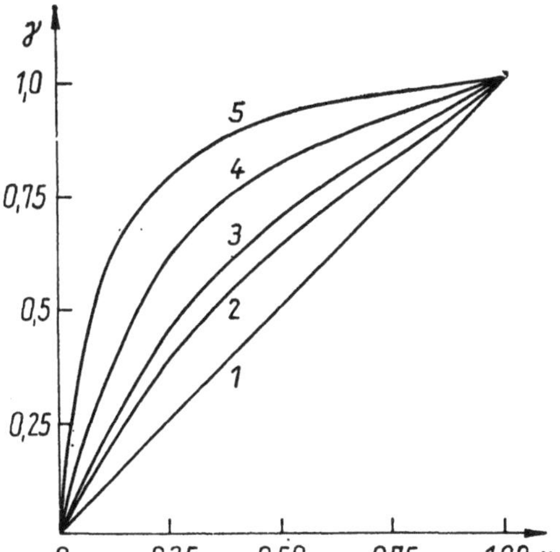

Fig. 6: Values of concentrations γ and y for the asymptotic stage of the frontal process.

$\varepsilon = 0.4$; $x = K_\lambda = 1$; $\Delta P = 1/6$; $P = 0.1$;
h: 1—0; 2—1; 3—5; 4—10; 5—∞.

ence upon the current sorption rate γ whereas actually it can be constant as well.

In the experimental evaluation of $\beta_E{}^*$ dependence upon γ, for example by employing front adsorption isoplan it is necessary to take into consideration the difference between C and the subsidiary variable C_1 as ZELDOVICH's rate applied in this case holds for a and c_1 rather than for c. Fig. 6 ad hoc shows dependences of γ current values upon y under various h. As it is seen from the figure ZELDOVICH's ratio holds only for $h \to \infty$, when longitudinal effect influences become insignificant. When the final flow rates $v\gamma > y$, curve $\gamma = \gamma(y)$ is convex and when $h \to 0$ the curve coincides with the isotherm.

5. Conclusions

A model of longitudinal mass-transfer and mass-exchange kinetics in granular bed, in which their mutual influence is taken into account is built.

It is shown that longitudinal effects stem from both turbulent fluctuations in carrier flow and sorbing substance longitudinal diffusion in the solid phase and depend upon mass-exchange intensity. The proposed equation of longitudinal mass-transfer differs from the known ones in diffusion members depending upon the absorbed substance concentration distribution over the bed length.

Based upon the assumption of turbulent mechanism of substance supply to the grain surface an expression for external mass-exchange coefficient evaluation was obtained. Equations for the asymptotic stage of both intradiffusion and total kinetics were found. The method of the approximate registratiion the longitudinal effect in the reduced mass-exchange coefficient permits to reduce the mathematical description of the sorption process to the system of differential equations with partial derivatives of hyperbolic type for the average γ according to the bed section and certain subsidiary functions connected with concentrations in both phases.

It was shown that with flow rate growth or grain diameter diminution the reduced mass-exchange coefficient increases. The relative part of longitudinal effects decreases with the growth of both grain diameter and flow rate.

When the sorption front is formed reduced and effective mass-exchange coefficients decrease with sorption rate growth, however, if the front is not formed $\beta_E{}^*$ can pass over the minimum with the sorption rate growth increasing in the region of great values of γ.

Intradiffusional and effective mass-exchange coefficients and the contribution of longitudinal effects connected with longitudinal diffusion increase simultaneously with internal diffusion coefficient growth. In so doing the reduced mass-exchange coefficient in the region of low D_i increases up to the maximum and then falls.

6. Designation

x, t — coordinate and time from the beginning of the process
c, c_0, $\bar{c}^*_s$, $\bar{c}^*$ — current and characteristic concentrations in the flow and equilibrium with superficial and integral sorption rates averaged according to the bed section
a_i, $\bar{a}_s$, $\bar{a}$, a_∞, a_0 — local bed section, averaged superficially, averaged integrally, limiting and equilibrium sorption rate.
ε — bed porosity
v_{fl}, v, u, u_0 — velocities of fluctuations, flow and concentrating points of the sorption front in the case of $\bar{a}$ other than 0 and equal to 0, respectively.
D_e, D_i, D_E — molecular diffusion coefficients in the flow of internal diffusion and longitudinal effect coefficients with no mass-exchange
λ, λ_1 — turbulence scale and average length of sorbate molecule continuous movement in one direction;
t_{rel} — relaxation time;
β_e, β_i, β_E, $\beta_E{}^*$ — coefficients of external, internal, effective and reduced mass-exchange;
d, R — grain diameters and radii.
p_1, p_2 — probability of globules movement in direct and opposite direction;

$$\Delta p = p_1 - p_2$$

k_λ — coefficient of proportionality between λ and d;
q_{1x}, q_{2x} — turbulent flows in direct and opposite direction;
b — constant of Langmuir isotherm equation;

$$y = \frac{c}{c_0}\ ; \gamma = \frac{\bar{a}}{a_0}\ ; p = \frac{1}{bc_0}; \varrho = \frac{D_e}{\beta d^2}\ ; h = \frac{vd}{D_e};$$

$$\varkappa = \frac{D_i}{D_e}\left(\frac{d\bar{a}}{d\bar{c}^*}\right)_{\bar{a}=0}; \alpha = \frac{\lambda_1}{\lambda}$$

7. References

[1] Levenspiel, O. "*Chemical Reaction Engineering*", Wiley, New York 1965
[2] Aerov, M. È., Todes, O. M., "*Hydrodynamic and Thermal Foundations of Fixed and Fludized Bed Apparatous*", Chemistry, Moscow 1968
[3] Radushkewitch, L. V., DAN USSR **51**, 471 (1947)
[4] Turner, G. A., Chem. Eng. Sci. **7** 156 (1958), **10**, 14 (1958)
[5] Nikolajevski, V. N., Appl. Mechanics and Mathematics **23**, 1042 (1959)

Some problems of non-isothermal kinetics of sorption

(E. A. Ustinov, V. L. Kolin, I. N. Taganov)

Zusammenfassung:

Der Beitrag ist der Aufstellung mathematischer Modelle für die isotherme Sorptionskinetik in Sorbentien mit gleichmäßigverteilten aktiven Zentren gewidmet. Es werden nichtlineare Differentialgleichungen für den Stoff- und Wärmetransport abgeleitet, die gleichzeitig durch Wärmeleitung und sorbierte Moleküle verursacht wird.

Für die Koeffizienten der Wärme- und Stoffleitung, den äußeren und inneren Wärme- und Stoffaustausch wird die explizite Form ihrer Abhängigkeit von Parametern der Isotherme und der Struktur des Sorbents sowie des strömungstechnischen Zustandes ermittelt. Asymptotische Lösungen wurden für spezielle Grenzfälle erhalten, die wichtigsten Gesetzmäßigkeiten der Sorptionskinetik werden analysiert.

1. Introduction

The models of the kinetics of sorption that are usually employed in the circulation for various processes of mass-transfer may be reduced to the basic concepts of the diffusional nature of the isothermal transfer of the sorbate through the solid phase and the border layer. However, the coefficients of the mass-transfer or of the inner diffusion, derived in accordance with the established models (e.g. the isoplanes for the asymptotic stage of the frontal sorption), depend, in most cases, on the value of the sorption.

This dependence of the kinetic constants (including those constant in the physical sense) on the value of the sorption shows the imperfection of the physical premises on which the kinetics equations are based and also shows these equations to be of limited use.

This circumstange undermined the scientific value of the models employed and also renders difficult the engineering calculations for the mass-transfer processes, as it calls for special methods. Forthis reason the development of more adequate models for the kinetics of sorption has become quite an urgent matter.

2. The physical model

In order to establish some basic laws of the kinetics let us employ the most simple model of the sorbent with the active centers of. the same ability energetically and evenly distributed throughout the bulk of the sorbent, each center being able to absorb only one molecule. The sorbate is transfered inside the grains by the unattached molecules through the macropores and also by the diffusion of these molecules through the solid phase. Each molecule can move only as far as the next active center, overcoming while it moves a potential bar that differs in general from the work of exit of the molecule from the solid phase.

The sorption results from a large volume of inert gas moving through the granular bed. The sorbate is transfered to the grains' surface by molecular diffusion. The conditions of movement of the carrier-gas will be described through the model for ideal displacement, while in the time intervals equal to the period of relaxation t_{rel}, in the areas of the granular bed, which are equal in size to the value of the turbulency scale λ_{tur}, there occurs an instantaneons stirring up of the inert gas flow. The deviations of the temperature and the concentration of the sorbate inside the singled-out areas in the flow are smoothed with each stirring up. We will stipulate for the future that the value of $\frac{\lambda_{tur}}{t_{rel}}$ is close to that of the flow velocity v, while the turbulency scale is about equal to the diameter d of the grains of the sorbent. The coefficient longitudinal transfer D_{eff} for this model is equal to $0.5vd$, which is in agreement with various experimental data [1—4].

The heat-transfer is carried out by the thermal conductivity of the carrier gas near the grains' surface and in the macropores and also by the thermal conductivity of the sorbent and by the molecules absorbed, which diffuse through the solid phase.

3. The basic equations

Let us begin with the balanced system of sorbent-sorbate. Further let us assume the energy of the absorbed molecule to be equal to $\frac{3}{2} kT - u_0$, where u_0 is independent of temperature and is equal to the energy during the absorption of one molecule of a uniatomic gas. From the First Law of Thermodynamics, the entropy of an ideal uniatomic gas is

$$S_1 = n_c k \ln T^{3/2}/KC \tag{1}$$

where:

n_c — is the number of molecules of the gas;
K — constant of the integration, which is independent of temperature and volume of the gas.

Then the change in the entropy of the gas, due to the adsorption of one molecule in the closed system "sorbent-sorbate", is

$$\Delta S_1 = -k \ln T^{3/2}/KC \tag{2}$$

The entropy of the absorbed substance

$$S_2 = k \ln W + n_a k \ln T^{3/2}/KC \tag{3}$$

where $W = \dfrac{N!}{(N - n_a)!n_a}$; the change due to the absorption of one molecule:

$$\Delta S_2 = k \ln \frac{1-\theta}{\theta} + k \ln T^{3/2}/KC \tag{4}$$

$\theta = n_a/N = a/a_\infty$ — the share of the occupied active centers. The total change in the entropy of the "sorbent-sorbate" system, due to the adsorption of one molecule:

$$\Delta S = \Delta S_1 + \Delta S_2 = k \ln \frac{1-\theta}{\theta}. \tag{5}$$

For the state of equilibrium, the change in the value of free energy F due to the transfer of a molecule from one phase into another

$$\Delta F = \Delta u - T\,\Delta S = 0. \tag{6}$$

As $\Delta u = -u_0$ and considering (5) we get from (6)

$$\theta = \frac{bc}{1 - bc}; \quad ba_\infty = e^{\frac{u_0}{kT}}. \tag{7}$$

We assumed here that as $T \to \infty$ $a \to c$, therefore $K = 1/a_\infty$.

The form of the equation (7) is closely similar to that of the LANGMUIR equation and it describes sorption not on a smooth surface but inside the bulk of sorbent.

In order to derive the equations for the transfer inside the solid phase we assume that the active centers are densely packed three-dimensional potential boxes, inside which the molecules have three degrees of freedom and obey MAXWELL's law of the distribution by velocities. In this case, the probability of the molecule overcoming a potential barrier with the height of $\Delta u = u_0 - u_i$ between two neighbouring potential boxes, in a unit of time is equal to:

$$\lambda_{mi}^2 \frac{1}{\lambda_{mi}^3} \left(\frac{m}{2\pi kT}\right)^{1/2} \int\limits_{\sqrt{\frac{2\Delta u}{m}}}^{\infty} v e^{-\frac{mv^2}{2kT}}\, dv = -\frac{1}{\lambda_{mi}} \sqrt{\frac{kT}{2\pi m}}\, e^{-\frac{\Delta u}{kT}}.$$

This allows for the area of potential boxes to be equal to λ_{mi}^2, and for "the concentration" of molecules per volume of one occupied potential box to be equal to $1/\lambda_{mi}^3$.

Then the mean time of the continuous stay of a molecule on the active center is:

$$t_{Pb} = \frac{1}{6} \lambda_{mi} \sqrt{\frac{2\pi m}{kT}}\, e^{\frac{\Delta u}{kT}}. \tag{8}$$

The flow of the molecules being desorbed off the unit surface of the sorbent is:

$$q_d = \frac{1}{\lambda_{mi}^3} \sqrt{\frac{kT_s}{2\pi m}}\, \theta_s e^{-\frac{u_0}{kT_s}} = a_s \sqrt{\frac{kT_s}{2\pi m}}\, e^{-\frac{u_0}{kT_s}}. \tag{9}$$

The number of molecules being absorbed in a unit time on the unit surface of the sorbent is, evidently, equal to:

$$q_a = c_s \sqrt{\frac{kT}{2\pi m}} (1 - \theta_s). \tag{10}$$

Hence, when there is equilibrium, from (9), (10) we again derive (7). Inside the solid phase the mass flow along e.g. the axis x is:

$$q_{cmi_x} = \frac{1}{6} \lambda_{mi} \left(\frac{a}{t_{Pb}}\right)_{x-\frac{\lambda_{mi}}{2}} \left(1 - \theta_{x+\frac{\lambda_{mi}}{2}}\right) - \frac{1}{6} \lambda_{mi} \left(\frac{a}{t_{Pb}}\right)_{x+\frac{\lambda_{mi}}{2}} \left(1 - \theta_{x-\frac{\lambda_{mi}}{2}}\right),$$

or, going over from differences to differentials and further generalizing for the three-dimensional case we get:

$$q_{cmi} = -D_{mi}[\vec{\nabla} a + a(1 - \theta) \vec{\nabla} \ln \alpha], \; D_{mi} = \frac{\lambda_{mi}^2}{6t_{Pb}} \tag{11}$$

In this equation $\alpha = t_{Pb}(T_0)/t_{Pb}(T)$; D_{mi} — coefficient of inner diffusion.
When $T = T_0 = \text{const}$ $\alpha = 1$ and the expression (11) becomes linear and coincides in its form with FICK's law. Also in general case the mass diffusion flow is proportional to the gradients of concentration and of temperature.

The mean energy transferred by each molecule overcoming the potential barrier ΔU

$$\bar{E} = \frac{\int\limits_0^\infty dv_r \int\limits_{\sqrt{\frac{\Delta u}{m}}}^\infty \left(\frac{mv^2}{2} - u_0\right) v_x e^{-\frac{mv^2}{2kT}} dv_x}{\int\limits_0^\infty dv_r \int\limits_{\sqrt{\frac{\Delta u}{m}}}^\infty v_x e^{-\frac{mv^2}{2kT}} dv_x} = 2kT - u_1. \tag{12}$$

Then for the thermal flow due to the sorbate transfer we can write as follows:

$$q_{Emi}^{(a)} = -(2kT - u_1) D_{mi} [\vec{\nabla} a + a(1 - \theta) \vec{\nabla} \ln \alpha(2kT - u_1)] \tag{13}$$

or

$$q_{Emi}^{(a)} = (2kT - u_1) q_{cmi} - 2D_{mi} a(1 - \theta) \vec{\nabla} kT. \tag{14}$$

The heat transfer inside the solid phase of the sorbent is carried out not only by the molecules migrating over the active centers, but also by the heat-conductivity of the sorbent itself, which represents a component of the total heat flow:

$$q_{Emi}^{(s)} = -\frac{Ha}{k} K_{T^{(s)}} \vec{\nabla} kT. \tag{15}$$

Since as a rule $\Delta u < u_0$ then for the free molecules inside the sorbent, which make for the sorbate transfer and are in equilibrium with the molecules localized over the active centers, the concentration of the above-mentioned free molecules may be considerably higher than the initiate concentration of the sorbate in the carrier-flow. Hence it folllows that the carrier-flow contributed to by the sorbate transfer inside the sorbent may be considerable. Otherwise, when $\Delta u = u_0$, it could be expected that $q^{(a)}_{Emi} \ll q^{(s)}_{Emi}$.

In order to deduce the equations (11), (14) we assumed that each molecule can move only to the adjacent unoccupied active center. Otherwise the molecule bounces off that side of the potential box which borders on the occupied active center.

Other models of the transfer process are, however, possible. The molecule hitting on the occupied active center can, with a various degree of probability, bounce off, or knock out the molecule already lodging in the center, or these molecules can together form an unstable complex, and when this comes apart, one of the molecules will move with equal probability to each of the adjacent unoccupied active center and so on.

Let us take for example a rather likely case, when the molecule with the energy $> \Delta u$ on hitting an occupied active center knocks out with the probability of 1 the molecule already lodged there, exchanging the impulse with this molecule. Then

$$q_{cmi_x} = \frac{\lambda_{mi} a_\infty}{6 t_{Pb}} \left[\left(1 - \theta_{x + \frac{1}{2}\lambda_{mi}}\right) \sum_{i=0}^{\infty} \alpha_{x - \frac{2i+1}{2}\lambda_{mi}} \prod_{j=0}^{i} \theta_{x - \frac{2j+1}{2}\lambda_{mi}} - (\alpha\theta)_{x + \frac{1}{2}\lambda_{mi}} \right]. \quad (16)$$

Expanding the functions appearing in (16) into power series and ignoring the derivatives above the first order and also products of these derivatives, then, after some simple transformations, we obtain:

$$q_{cmi} = -D_{mi}[\vec{\nabla} a + a(1-\theta)\vec{\nabla} \ln \alpha], \qquad D_{mi} = \frac{\lambda^2_{mi}}{6 t_{Pb}} \frac{1}{(1-\theta)^2}. \quad (17)$$

The expression (17) concurs with (11), but the coefficient of inner diffusion in (17) depends on the value of sorption and with the rise in θ the coefficient D_{mi} rises up to ∞ (when $\theta = 1$). This conclusion is obvious, as with the rise in the value of sorption there also occurs the rise in probability of the formation of chaines out of the adjacent occupied active centers. As each dislodged molecule continues on the way of its fore-runner, the effective free-path of the molecules and, consequently, the coefficient of inner diffusion will rise with θ.

For the rate of desorption we can in a similar way find:

$$q_d = \sqrt{\frac{kT_s}{2\pi m}}\, a_\infty e^{-\frac{u_0}{kT_s}} \sum_{i=0}^{\infty} \alpha_{sx - \frac{2i+1}{2}\lambda_{mi}} \prod_{j=0}^{i} \theta_{x - \frac{2j+1}{2}\lambda_{mi}}.$$

If we ignore the values of second order of smallness we obtain

$$q_d = \sqrt{\frac{kT_s}{2\pi m}}\, e^{-\frac{u_0}{kT_s}} \frac{a_s}{1-\theta_s} = \sqrt{\frac{kT_s}{2\pi m}}\, C^*(a_s). \quad (18)$$

As for the model under consideration, each molecule is absorbed from the gaseous phase, which hits the sorbent surface, then in case of equilibrium taking (18) into account, we again, as should be expected, obtain the equation for isotherm (7).

For the heat-flow due to the sorbate transfer inside the solid phase we shall come to an expression coinciding with (13), in which the coefficient of inner diffusion is derived from (17).

Thus it can be expected that for the various modes of sorbate transfer the form of equations (11), (17) for the mass and heat flows remains the same, but for these cases the dependence of the coefficient of inner sorption on the value of sorption conforms to the different laws. In the future the specific mode of transfer may be chosen by way of eliminating the different kinetic hypotheses by the employment of the solutions of the inverse problems.

Let us now examine the transfer of the sorbate in the macropores, which function as the transportation channels. We will deem the macropores accessible to the molecules of the carrier-gas, while the concentration of the gas in the macropores as well as outside the grains of the sorbent will be considerably higher than the concentration of the sorbate. Therefore we will ignore the collisions between the sorbate molecules and in the future we will consider the heat transfer in the macropores to be connected only with the thermoconductivity of the carrier-gas.

In order to derive the equations for the transfer in the gas it is necessary to employ the kinetic equation by Boltzmann [5]

$$\frac{\partial Cf}{\partial t} + \vec{v}\vec{\nabla} Cf = \sigma C^2 \int |v_1 - v| \, (f'f_1' - ff_1) \, d^3v. \tag{19}$$

This equation does not account for the field of the external forces, and the section of the collision σ is considered to be constant; f — the distribution function for the molecules by their velocities. The existing ways of solving this equation are as a rule rather labourconsuming, e.g. the Chapman-Enscoge method [6] gives the solution the first approximation to f as a system of differential equations, the solution of the system being, in itself, rather difficult a task. The approximate method, introduced in [7] is based on the assumption that the free-path time of the molecules is independent of their velocities, while in fact the free-path time decreases with the increase in velocity. Therefore we will make use here of the following concept. We assume that immediately after the molecules collide with each other or with the pore surfaces, or during desorption, the distribution of the molecules by their velocities obeys Maxwell's law. This assumption is quite natural because as the result of the collisions of the molecules the system tends rapidly to attain the equilibrium, if the system has been hitherto driven out of it. We assume the number of collisions of a molecule per unit of time $\nu(v)$ to be approximately equal to that for the equilibrium function of distribution. Then for the collision integral in (19) we obtain

$$\sigma C^2 \int |v_1 - v| \, (f'f_1' - ff_1) \, d^3v \cong \nu(Cf_0 - Cf) \tag{20}$$

where

$$f_0 = \left(\frac{m}{2\pi kT}\right)^{3/2} e^{-\frac{m(v-V)^2}{2kT}} \tag{21}$$

$$\nu = \sigma C \sqrt{\frac{2kT}{\pi m}} \, \varrho; \quad \varrho = e^{-z^2} + \sqrt{\pi}\left(z + \frac{1}{2z}\right) \operatorname{erfc} z, \quad r^2 = \frac{m(v - V)^2}{2kT}. \tag{22}$$

So in the expression (20) the functions f_1, f_1', f' are replaced by zero approximation f_0. And, finally, the approximation to BOLTZMANN's equation assumes the following aspect

$$\frac{\partial Cf}{dt} + \vec{v}\vec{\nabla} Cf + \nu(Cf - Cf_0) = 0. \tag{23}$$

The first approximation to f may be as a matter of fact again substituted into the collision integral and by this the second approximation can be obtained and so on. Here we will restrict ourselves to the first approximation.

If we rougly replace cf in (23) and also if we take into accoint the collisions of the molecules with the macropores' surface and the desorption of the absorbed molecules of the carrier-gas and of the sorbate, then we obtain, correspondingly, the following equations

$$\frac{\partial C_1 f_{10}}{\partial t} + \vec{v}\vec{\nabla} C_1 f_{10} + \nu_1(C_1 f_1 + C_1 f_{10}) = 0 \tag{24}$$

$$\frac{\partial C^* f_0}{\partial t} + \vec{v}\vec{\nabla} Cf_0 + \nu(Cf - C^* f_0) = 0 \tag{25}$$

where

$$f_{10} = \left(\frac{m_1}{2\pi kT}\right)^{3/2} \cdot e^{-\frac{m_1 v^2}{2kT}}; \quad f_0 = \left(\frac{m}{2\pi kT}\right)^{3/2} e^{-\frac{mv^2}{2kT}} \tag{26}$$

$$\nu_1 = \nu_{11} + \nu_{ma}, \qquad \nu = \nu_{01} + \nu_{ma} \tag{27}$$

$$\nu_{11} = \sigma_{11} C_1 \sqrt{\frac{2kT}{\pi m_1}} \left[e^{-z_1^2} + \sqrt{\pi}\left(z_1 + \frac{1}{2z_1}\right) \operatorname{erf} z_1 \right], \quad z_2 = \sqrt{\frac{m_1}{2kT}}\, v \tag{28}$$

$$\nu_{ma} = \frac{1}{4} \frac{s_{ma}}{V_{ma}} v, \qquad \nu_{01} = \frac{\sigma_{01}}{\sigma_{11}} \nu_{11}. \tag{29}$$

Here ν_{11} and ν_{01} are the frequencies of collisions with the molecules of the carrier-gas of the isolated molecules of the carrier itself and of the sorbate; ν_{ma} is the frequency of the collisions of the molecules either of the sorbate or of the carrier with the macropores' surface, of which the specific surface and volume are, accordingly, s_{ma} and V_{ma}. As after the collisions of the molecules with immobile surface of the macropores their distribution function becomes symmetrical, the mean velocity $V = Svf_0 d^3v$ is for the zero approximations (26) assumed to be equal to zero.

For the impulse flow we can derive from (24) the following:

$$q_{mvma} = P + m_1 \frac{16}{3\sigma_{11}\sqrt{2m_1}} B_{cma}^{(g)} \frac{\partial \sqrt{kT}}{\partial t} \tag{30}$$

where

$$B^{(g)}_{cma} = \int_0^\infty \frac{z_1^4 \left(\frac{5}{2} - z_1^2\right) e^{-z_1^2}\, dz_1}{\varrho + \frac{\sqrt{\pi}}{4} \frac{S_{ma}}{V_{ma}} \frac{1}{\sigma_{11} C_1} z_1}.$$

The second member in the right part of (30) may be ignored. As the mean velocity V and, consequently, its changes with time are in zero approximation equal to zero, then f should be nearly constant and equal to the pressure P_0 in the carrier-flow. Then C_1 in the expressions (24) and (28) may be replaced by P_0/kT. In this case we obtain the following expression for the heat flow in the macropores:

$$q_{Ema} = C_1 \int \vec{v} \frac{mv^2}{2} f_1 d^3v = -\frac{P_0}{kT} K_{Tma} \vec{V} kT \tag{31}$$

where the coefficient of thermoconductivity

$$K_{Tma} = \frac{8}{3\sigma_{11} \sqrt{2m_1}} B^{(g)}_{Ema} \frac{(kT)^{3/2}}{P_0} \tag{32}$$

while

$$B^{(g)}_{Ema} = \int_0^\infty \frac{z_1^6 \left(z_1^2 - \frac{5}{2}\right) e^{-z_1^2} dz_1}{p + \frac{\sqrt{\pi}}{4\sigma_{11}} \frac{s_{ma}}{V_{ma}} \frac{kT}{P_0} z_1}. \tag{33}$$

But from (33) it is evident that the integral $B^{(g)}_{Ema}$ depends on temperature and pressure F_0. As the mean number of collisions of the gas molecules with each other $\overline{\nu_{11}}$ and the mean number of collisions of the molecules with the macropores surface $\overline{\nu_{ma}}$ are, correspondingly, equal to $C_1 \sqrt{2}\, \sigma_{11} \bar{v}$ and $\bar{v}/\lambda_{ma}$, where $\lambda_{ma} = 4T_{ma}/S_{ma}$, then if $P_0 \lambda_{ma} \sqrt{2}\, \sigma_{11}/kT \gg 1$, $B^{(g)}_{Ema} = \text{const}$ and the coefficient of thermoconductivity $K_{Tma} \sim \sqrt{kT}$ and is independent of the pressure P_0, though in this case tze transfer is carried out only by collisions between the molecules of the carrier-gas.

When $P_0 \bar{n}_{ma} \sqrt{2}\, \sigma_{11}/kT \ll 1$ then the thermoconductivity coefficient is

$$K_{Tma} = \frac{4}{3} \frac{\lambda_{ma}}{\sqrt{2\pi m_1}} P_0 \frac{1}{\sqrt{kT}}$$

and is proportional to pressure and specific volume of the macropores and inversely proportional to the specific surface.

For the sorbate flow we obtain from (25) the following expression:

$$q_{cma} = \int vCf d_3v = -\frac{8(kT)^{3/2}}{3\sqrt{2m_1}\, \sigma_{01} P_0} \left[B_{cma} \vec{V} C^* - B'_{cma} \frac{C^*}{kT} \vec{V} kT\right] \tag{34}$$

where

$$B_{cma} = \int_0^\infty \frac{z_1^2 z^2 e^{-z^2} dz}{\varrho + \frac{\sqrt{\pi}}{\lambda_{ma}\sigma_{01}} \cdot \frac{kT}{P_0} z_1}. \tag{35}$$

$$B'_{cma} = \int_0^\infty \frac{z_1^2 z^2 \left(\frac{3}{2} - z^2\right) e^{-z^2} dz}{\varrho + \frac{\sqrt{\pi}}{\lambda_{ma}\sigma_{01}} \frac{kT}{P_0} z_1} \tag{36}$$

The concentration C^* in (34) is in equilibrium with the value of sorption $\dot{a}$ and may be derived from the isotherm equation (7). As, usually, the value $f \frac{u_0}{kT} \gg 1$, we may ignore the second member in the right part of the equation (34) and the expression for the sorbate flow assumed the following form:

$$q_{cma} = -D_{ma}[\vec{\nabla} a + (1 - \theta) \vec{\nabla} \ln \alpha_1] \tag{37}$$

where

$$\alpha_1 = e^{-\frac{u_0}{kT}}, \; D_{ma} = \frac{8(kT)^{3/2}}{3\sigma_{01} \sqrt{2m_1} P_0} B_{cma} e^{-\frac{u_0}{kT}} \frac{1}{(1 - \theta)^2}. \tag{38}$$

The form of the equation (37) is similar to that of (11), the coefficient of innerdiffusion in macropores depends on the value of sorption. When $P_0 \lambda_{ma} \sqrt{2}\, \sigma_{01}/kT \gg 1$, $B_{cma} = \text{const}$ and D_{ma} is inversely proportional to the pressure P_0. With the decrease in pressure the D_{ma} increases and tends when $P_0 \lambda_{ma} \sqrt{2}\, \sigma_{01}/kT \ll 1$ to the limit, that is:

$$D_{ma} = \frac{4}{3} \sqrt{\frac{kT}{2\pi m}} \lambda_{ma} e^{-\frac{u_0}{kT}} \frac{1}{(1 - \theta)^2}. \tag{39}$$

The expression (39) is similar in form to (17) for the sorbate flow inside the solid phase, and the coefficients of diffusion D_{ma} and D_{mi} depend on the value of sorption by the same law. The ratio of the coefficients

$$\frac{D_{mi}}{D_{ma}} = \frac{3}{4} \frac{\lambda_{mi}}{\lambda_{ma}} e^{\frac{u_1}{kT}}. \tag{40}$$

The ratio of the length of the free-paths of the sorbate molecules inside the solid phase to that inside the macropores $\lambda_{mi}/\lambda_{ma}$ is, as a rule, less than 1, but the exponential factor in the right part of (40) can be of substantial value. Hence the rate of diffusion of the adsorbed molecules can be exceedingly higher than the rate of diffusion of the molecules of sorbate in the macropores. Though, apparently, as a rule this circumstance occurs. Otherwise, as the length of the diffusional zone in the macropores is always more than near the grains' surface and the coefficient of diffusion of the sorbate is always small, it could be expected that the process of sorption will always take place in the inner-diffusional area. Then the total flows throughout the grains' volume may be put down in the following way:

$$q_{ci} = -D_i \vec{V} a - \left(D_{mi} + \frac{u_0}{\Delta u} D_{ma}\right) \frac{\Delta u}{(kT)^2} a(1-\theta) \vec{V} kT \qquad (41)$$

$$q_{Ei} = -(2kT - u_1) D_{mi} \left[\vec{V} a + \frac{\Delta u}{(kT)^2} a(1-\theta) \vec{V} kT\right] - \left(\frac{H_a}{k} K_T^{(s)} + \frac{P_0}{kT} K_{Tma}\right) \vec{V} kT. \qquad (42)$$

Here we take it that $\frac{u_0}{kT_0} \gg 1$; $\frac{\Delta u}{kT_0} \gg 1$ and $\frac{\Delta kT}{kT} \ll 1$.

The equations of material and thermal balances evidently should appear in the following form:

$$\vec{V} \vec{q}_{ci} + \frac{\partial a}{\partial t} = 0, \vec{V} \vec{q}_{Ei} + \frac{\partial}{\partial t}\left[H_a T + \left(\frac{3}{2} kT - u_0\right) a\right] = 0. \qquad (43)$$

The system (41)—(43) describes the space-time distribution of temperatures T and value sorption a inside the grain of the sorbent. It should be noted as a characteristic feature that the solution $kT = kT_0 = \text{const}$ does not satisfy the system (41)—(43). Therefore the problem of the inner diffusion of the sorbate should always be considered to be non-isothermal.

In order to deduce the equations for the transfer close to the grains' surface we assume that Stefan's flow does not add to the total heat-transfer, because when there is a large number of the grains in the bed the thermal radiation is in equilibrium with the heated surface of the grains. Hence the heat-transfer is conducted only by the thermo-conductivity of the inert carrier.

In order to obtain the solution for the problem stated, it is essential to begin with the deduction of the equation of transfer for the carrier-gas and after that to apply the distribution thus obtained of the temperatures and mean velocities of the molecules to the deduction of the expression for the sorbate transfer.

Boltzmann's equation for the carrier-gas appears in the following forms:

$$\frac{\partial C_1 f_1}{\partial t} + \vec{v} \vec{V} C_1 f_1 + \nu_{11}(C_1 f_1 - C_1 f_{10}) = 0 \qquad (44)$$

where

$$f_{10} = \left(\frac{m_1}{2\pi kT}\right)^{3/2} e^{-\frac{m(v-V)^2}{2kT}}$$

$$\nu_{11} = \sigma_{11} C_1 \sqrt{\frac{2kT}{\pi m_1}} \varrho(z_1), \quad z_1^2 = \frac{m_1(v - V)^2}{2kT}.$$

Here the velocity V is not generally equal to zero because otherwise we would obtain three equations for the transfer (mass-, heat- and quantity of motion-transfer) for only two variables, C_1 u T. This assertion remains correct also in case of the mass-flow being equal to zero. Let us replace approximately $C_1 f_1$ under the derivative symbol in (44) by the zero approximation.

$$\frac{\partial C_1}{\partial t} + \frac{1}{kT} \vec{V}\vec{V} P + C_1 \vec{V}\vec{V} - \frac{C_1}{kT} \vec{V}\vec{V} kT = 0 \tag{45}$$

$$\frac{\partial m\vec{V}}{\partial t} + \frac{1}{C_1} \vec{V} P + (\vec{V}\vec{V})\, m\vec{V} = 0 \tag{46}$$

$$\frac{\partial kT}{\partial t} + \frac{2}{3} kT \vec{V}\vec{V} + \vec{V}\vec{V} kT = 0 \tag{47}$$

after some necessary reductions we obtain

$$f_1 = f_{10} + \frac{1}{kT} f_{10} \frac{1}{\nu_{11}} \left[\left(\frac{5}{2} - z_1^2\right) \vec{W}\vec{V} kT + m_1 \vec{W}(\vec{W}\vec{V})\,\vec{V} - \frac{1}{3} m_1 W^2 \vec{V}\vec{V}\right] \tag{48}$$

where $\vec{W} = \vec{v} - \vec{V}$. Then, considering that usually $V \ll \sqrt{\dfrac{2kT}{m_1}}$, we can obtain from (48) the following expressions for the flows

$$\vec{q}^{\,(g)}_{Ce} = C_1 \vec{V} + \frac{16}{3\sigma_{11}\sqrt{2m_1}} B^{(g)}_{Ce} \vec{V} \sqrt{kT} \tag{49}$$

$$\vec{q}_{Ee} = C_1 \vec{V} \frac{5}{2} kT - \frac{16}{3\sigma_{11}\sqrt{2m_1}} B_{Ee} kT \vec{V} \sqrt{kT} \tag{50}$$

or, by eliminating $C_1\vec{V}$ from (41), (42)

$$\vec{q}_{Ee} = \frac{5}{2} kT \vec{q}^{\,(g)}_{Ce} - \frac{16}{3\sigma_{11}\sqrt{2m_1}} \left(\frac{5}{2} B^{(g)}_{Ce} + B_{Ee}\right) kT \vec{V} \sqrt{kT} \tag{51}$$

where

$$B^{(g)}_{Ce} = \int\limits_0^\infty \frac{z_1^4 \left(\frac{5}{2} - z_1^2\right) dz_1}{1 + \sqrt{\pi}\left(z_1 + \frac{1}{2z_1}\right) e^{z_1^2} \operatorname{erf} z_1}$$

$$B_{Ee} = \int\limits_0^\infty \frac{z_1^6 \left(z_1^2 - \frac{5}{2}\right) dz_1}{1 + \sqrt{\pi}\left(z_1 + \frac{1}{2z_1}\right) e^{z_1^2} \operatorname{erf} z_1}.$$

For the impulse flows, if we ignore the viscosity friction in the carrier-flow, from (48) we get

$$q_{mVij} = \begin{cases} P + mV_{x_i} q^{(g)}_{Ce_i}, & i = j \\ mV_{x_j} q^{(g)}_{Ce_i}, & i \neq j \end{cases} ; \quad i, j = 1, 2, 3 \tag{52}$$

where q_{mVij} — the impulse flow along the axis x_i, due to the transfer of the molecules along the axis x_j; $q^{(g)}_{Ce_i}$ mass-flow along the axis x_i. Further, if we assume $\frac{\partial V}{\partial t} \approx 0$, from (52) we obtain $P = P_0 = \text{const}$ and, consequently, $C_1 = P_0/kT$.

Then, from the equations for the material and thermal balances

$$\vec{V}\vec{q}_{Ee} + \frac{\partial C_1 \frac{3}{2} kT}{\partial t} = 0, \qquad \vec{V}\vec{q}^{(g)}_{Ce} + \frac{\partial C_1}{\partial t} = 0 \tag{53}$$

if follows that

$$\vec{V}\vec{q}_{Ee} = 0 \tag{54}$$

$$\frac{\partial kT}{\partial t} = \frac{1}{P_0} (kT)^2 \vec{V}\vec{q}^{(g)}_{ce}. \tag{55}$$

Excluding $\vec{q}^{(g)}_{Ce}$ from (51), (55) we obtain the following equation of heat-conductivity:

$$\frac{\partial kT}{\partial t} + \frac{2}{5} \frac{q_{Fe}}{P_0} \vec{V} kT = 2K_T \sqrt{kT}\, V^2 \sqrt{kT} \tag{56}$$

where the coefficient of thermo-conductivity for the gas is:

$$K_T = \frac{8}{3\sigma_{11}\sqrt{2m_1}} \left(B^{(g)}_{ce} + \frac{2}{5} B_{Ee}\right) \frac{(kT)^{3/2}}{P_0}. \tag{57}$$

Equation (56) differs from the more common equation for heat-conductivity by Fourier in that it contains a member proportional to the gradient of temperatures and to intensity of the thermal flux. For the sorbate, diffusing in the carrier-gas, Boltzmann's equation assumes the following form:

$$\frac{\partial Cf}{\partial t} + \vec{v}\vec{V}\, Cf + \nu_{10}(Cf - Cf_0) = 0 \tag{58}$$

where

$$f_0 = \left(\frac{m}{2\pi kT}\right)^{3/2} e^{-\frac{m(v-V)^2}{2kT}}, \quad \nu_{10} = \frac{\sigma_{01}}{\sigma_{11}} \nu_{11}$$

then

$$f = f_0 - \frac{1}{c\nu_{01}} f_0 \left(\frac{\partial C}{\partial t} + \vec{v}\vec{V}C\right) + \frac{1}{\nu_{01} kT} f_0 \left(\frac{3}{2} - z^2\right)\left(\frac{\partial kT}{\partial t} + \vec{v}\vec{V} kT\right) \tag{59}$$

and, further, for the diffusional flow of the sorbate:

$$q_{ce} = CV - \frac{8}{3\sigma_{01}\sqrt{2m_1}} B_{ce} \frac{(kT)^{3/2}}{P_0} \vec{V} C - \frac{8}{3\sigma_{01}\sqrt{2m_1}} B'_{ce} \frac{1}{P_0} C \sqrt{kT}\, \vec{V} kT \tag{60}$$

where

$$B_{ce} = \int_0^\infty \frac{1}{\varrho} z_1{}^2 z^2 e^{-z^2} dz, \quad B'_{Ce} = \int_0^\infty \frac{1}{\varrho} z_1{}^2 z^2 \left(z^2 - \frac{3}{2}\right) e^{-z^2} dz. \tag{61}$$

Replacing v in (60) by the expression for it from (50) we obtain:

$$q_{ce} = \frac{2}{5} \frac{q_{Ee}}{P_0} C - D_e \vec{V} C + K_T \eta \frac{C}{kT} \vec{V} kT. \tag{62}$$

The coefficients of the molecular diffusion D_e and η we may get from the following expressions

$$D_e = \frac{8}{3\sigma_{01}\sqrt{2m_1}} B_{ce} \frac{(kT)^{3/2}}{P_0} = K_T \frac{\frac{\sigma_{11}}{\sigma_{01}} B_{ce}}{B^{(g)}_{Ce} + \frac{2}{5} B_{Ee}} \tag{63}$$

$$\eta = \frac{\frac{2}{5} B_{Ee} - \frac{\sigma_{11}}{\sigma_{01}} B'_{Ce}}{B^{(g)}_{Ce} + \frac{2}{5} B_{Ee}} \tag{64}$$

as can be observed from (61), (63), (64); the coefficients of molecular diffusion D_e and η contained in (62) depend, if the other conditions are equal, on the effective section of collisions σ_{01} of the sorbate molecules with those of the carrier and on their mass ratio m/m_1.

The equation (62) differs from Fick's law and leads to the appearance of the constituents of the diffusional flow, which are proportionate to the heat flow and to the gradient of the temperature and which concur with them directionally.

Let us now consider the problem of the heat- and sorbate-transfer near the grains' surface with the assumption that the relaxation time t_{rel} is so small that $\delta_\tau = -\Delta kT_s \Big/ \left(\frac{\partial kT}{\partial r}\right)_R \ll R$ and $\delta_c = -C_s \Big/ \left(\frac{\partial C}{\partial r}\right)_R \ll R$. Then the problem with become one-dimensional and we have instead of (56):

$$\frac{\partial kT}{\partial t} + \frac{2}{5} \frac{q_E}{P_0} \frac{\partial kT}{\partial r} = 2K_T \frac{\partial^2 \sqrt{kT}}{\partial r^2} \sqrt{kT}. \tag{65}$$

In order to obtain an approximate solution of this problem we will employ the following method. When searching for function kT we will approximate near the grain surface by a polynomial of the second order the area $R \leqq r \leqq r_1$, and we will further stipulate that, at the point r_1, $kT = kT_0$ and $\dfrac{\partial kT}{\partial r} = 0$; and when $r = l$ then $kT = kT_s =$ const. during the relaxation time.

Then

$$kT = \begin{cases} kT_0 + (kT_s - kT_0)\left(\dfrac{r_1 - r}{r_1 - R}\right)^2, & R \leqq r \leqq r_1 \\ kT_0, & r_1 < r \end{cases} \tag{66}$$

as q_{Ee} is the function of time only and $q_{Ce}^{(g)}$ when $r = R$ is obviously equal to zero then it follows from (51), that:

$$q_{Ee} = \frac{5}{2}\,\frac{P_0}{kT_s}\,K_{T_s}\,\frac{\partial kT}{dr}\bigg|_R \tag{67}$$

or, if we take (66) into account

$$q_{Ee} = 5\,\frac{P_0}{kT_s}\,K_{T_s}\,\frac{\Delta kT_s}{r_1 - R} \tag{68}$$

as $q_{Ce}^{(g)} =$ const when $r > r_1$, then from (51) and with regard to (66), (68) it follows that

$$q_{Ce}^{(g)}\Big|_{r=\infty} = q_{Ce}^{(g)}\Big|_{r=r_1} = \frac{2}{5}\,\frac{q_{Ee}}{kT_0} = 2\,\frac{P_0}{kT_0kT_s}\,K_{T_s}\,\frac{\Delta kT_s}{r_1 - R}. \tag{69}$$

By integrating the equation for material balance (53) we obtain for the general case:

$$q_{Ce}^{(g)}\Big|_\infty + \frac{d}{dt}\int\limits_R^\infty (C_1 - C_{10})\,dr = 0 \tag{70}$$

as the pressure $P_0 = C_1kT$ is constant for all values of $r > R$ then the integral in (70) is:

$$\int\limits_R^\infty (C_2 - C_{10})\,dr = P_0\int\limits_R^{r_1}\left(\frac{1}{kT} - \frac{1}{kT_0}\right)dr$$

$$= -\frac{P_0}{kT_0}\left[1 - \sqrt{\frac{kT_0}{\Delta k\,T_s}}\ \text{arctg}\sqrt{\frac{\Delta kT_s}{kT_0}}\right](r_1 - R). \tag{71}$$

Then from (69), (70), (71) we can obtain the equation relatively to ·

$$\frac{d(r_1 - R)}{dt}\left[1 - \sqrt{\frac{kT_0}{\Delta kT_s}}\ \text{arctg}\sqrt{\frac{\Delta kT_0}{kT_0}}\right] = 2\,\frac{\Delta kT_s}{kT_s}\,K_{T_s}\,\frac{1}{r_1 - R} \tag{72}$$

and, after integrating (72) we get:

$$r_1 - R = 2\sqrt{3\,\frac{K_{T_s}}{\varkappa}\,t}, \qquad \varkappa = 3\,\frac{kT_s}{\Delta kT_s}\left[1 - \sqrt{\frac{kT_0}{\Delta kT_s}}\ \text{arctg}\sqrt{\frac{\Delta kT_s}{kT_0}}\right]. \tag{73}$$

whence

$$q_{Ee} = \frac{5}{2} \frac{P_0}{kT_s} \sqrt{\frac{\varkappa K_{T_s}}{3t}} \Delta kT_s \tag{74}$$

If $\Delta kT_s/kT_0 \ll 1$, $\varkappa \to 1$ (74) concurs with the solution to FICK's equation for heat-conductivity. And the precise solution for this case is

$$q_{Ee} = \frac{5}{2} \frac{P_0}{kT_0} \sqrt{\frac{K_T}{\pi t}} \Delta kT_s .$$

So it can be expected that the method introduced here of solution (65) leads to an insignificant error.

The mean integral heat flow of the unitary surface of the sorbent

$$\langle q_{Ee} \rangle = 5 \frac{P_0}{kT_s} \sqrt{\frac{\varkappa K_{T_s}}{3t_{\text{rel}}}} \Delta kT_s \tag{75}$$

is proportionate to $\sqrt{P_0}$ andfinversely proportionate to $\sqrt{t_{\text{rel}}}$. As it is stipulated that $t_{\text{rel}} = d/V$, we obtain for the heat-flow of the total grain surface per its unit volume:

$$\frac{6}{d} \langle q_{Ee} \rangle = 10 \frac{P_0}{kT_s} \sqrt{3\varkappa K_{T_s}}\, v^{0.5} d^{-1.5} (kT_s - kT_0) . \tag{76}$$

We will make a similar analysis of the equation for the sorbate transfer.

From the equation of the material balance follows that

$$q_{Ce}|_R = \frac{d}{dt} \int_R^\infty (C - C_0)\, dr . \tag{77}$$

Further we will look for the distribution of concentrations near the grain surface in the following form:

$$C - C_0 = \begin{cases} (C_s - C_0) \left(\dfrac{r_2 - r}{r_2 - R} \right)^2 , & R \leqq r \leqq r_2 \\ 0 , & r_2 < r . \end{cases} \tag{78}$$

Then

$$q_{Ces} = (C_s - C_0) \frac{1}{3} \frac{dr_2 - R}{dt} . \tag{79}$$

On the other hand, from (62) with regard to (66), (74) and (78) we get:

$$q_{ces} = (1 - \eta) \sqrt{\frac{\varkappa K_{T_s}}{3t}} \frac{\Delta kT_s}{kT_s} C_s + 2D_e (C_s - C_0) \frac{1}{r_2 - R} . \tag{80}$$

Excluding q_{ces} from (79), (80) and after integrating we get:

$$r_2 - R = \left(\mu + \sqrt{\mu^2 + 12 D_e} \right) \sqrt{t} , \qquad \mu = \frac{\Delta kT_s}{kT_s} \frac{C_s}{\Delta C_s} (1 - \eta) \sqrt{3 \varkappa K_{T_s}} \tag{81}$$

whence

$$q_{ces} = \frac{1}{6}\left(\mu + \sqrt{\mu^2 + 12D_e}\right)\frac{1}{\sqrt{t}}\,\Delta C_s. \tag{82}$$

If $\Delta kT_s = 0$, then if there is isothermal diffusion of the sorbate:

$$q_{ces} = \sqrt{\frac{D_e}{3t}}\,\Delta C_s.$$

For the mean flow of the sorbate to the total grains surface, as per its unit volume and with regard to (82), we obtain

$$\frac{6}{d}\langle q_{ces}\rangle = 2\left(\mu + \sqrt{\mu^2 + 12D_e}\right) v^{0,5} d^{-1,5}(C_s - C_0). \tag{83}$$

It is characteristic that the intensity of the diffusional flow of the sorbate depends, in general, not only upon the difference of concentrations in the core of the carrier-flow and on the surface ΔC_s, but also on the absolute value of the surface concentration C_s. The forms of the power dependences for the heat and concentration flows upon the carrier velocity and grain diameter d are the same. The gradient of temperature leads to the alteration of the diffusional flow of the sorbate by a component with a direction opposite to that of the gradient. So, if $\Delta C_s < 0$, the absolute value of the diffusional flow of the sorbate turns out to be less than for isothermal conditions when $\Delta kT_s > 0$ and higher when $\Delta kT_s < 0$.

The heat and difusional flows are connected through the equations for heat and mass-balance of the sorbate for the sorbent grain:

$$\frac{6}{d}\langle q_{Ee}\rangle = u_0\frac{d\bar{a}}{dt} - H_a\frac{d\overline{T}}{dt} - \frac{3}{2}\frac{d}{dt}\frac{3}{R^3}\int_0^R kTar^2\,dr \tag{84}$$

$$\frac{6}{d}\langle q_{ces}\rangle = -\frac{d\bar{a}}{dt}, \quad \bar{a} = \frac{3}{R^3}\int_0^R ar^2\,dr, \quad \overline{T} = \frac{3}{R^3}\int_0^R Tr^2\,dr. \tag{85}$$

The system (76), (83)—(85) should so be solved together with the equations for inner heat-mass-transfer (41), (42).

In particular, if sorption is not restricted to inner diffusion, then

$$\frac{6}{d}\langle q_{Ee}\rangle + \left(u_0 - \frac{3}{2}kT\right)\frac{6}{d}\langle v_{ces}\rangle = -\left(\frac{H_a}{R} + \frac{3}{2}a\right)\frac{dkT}{dt} \tag{86}$$

and in (83) for $\langle q_{ces}\rangle$ the surface concentration C_s is connected to a by the isotherm equation (7). The equations (85), (86) simply connect temperature and value of sorption for the specific supplementary conditions. Hence, if we exclude t from (85) we can write down the following:

$$\left(\frac{H_a}{k} + \frac{3}{2}a\right)\frac{dkT}{da} = \left(u_0 - \frac{3}{2}kT\right) + \frac{\langle q_{Ee}\rangle}{\langle q_{ces}\rangle}. \tag{87}$$

By integrating (87) and substituting $kT = kT(a)$ into (83) we obtain an expression for the rate of sorption $\frac{6}{d}\langle q_{ces}\rangle$ or for the coefficient of outer mass-exchange $\beta_e = \frac{6}{d}\langle q_{ces}\rangle/\Delta C_s$ as dependent on the value of sorption a.

During the adsorption if, e.g. at the moment, $t = 0$, $a = \Delta kT = \langle q_{Be}\rangle = 0$ then the temperature, as can be observed from (87), rises at the first stage. With the rise in T the heat flow $\langle q_{Be}\rangle > 0$ also increases and the sorbate flow $\langle q_{ces}\rangle < 0$ diminishes. Therefore the further increase in the temperature slows down, while it reaches the maximum, which is when $\langle q_{Be}\rangle = -\left(u_0 - \frac{3}{2}kT\right)\langle q_{ces}\rangle$, and subsequently the temperature decreases until it drops ot the initial value.

During desorption, when e.g. at the moment $t = 0$, $a = a_0$, $\Delta kT = 0$, the temperature at the first stage consequently passes through the minimum.

For the point of extremum kT it follows from (87) and with regard to (76), (83), that

$$\left|\frac{\Delta kT_m}{kT_0}\right| \leqq \frac{2}{5}\sqrt{\frac{D_e}{\varkappa K_T}}\,\frac{C_0}{C_{10}}\left(\frac{u_0}{kT_0} - \frac{3}{2}\right). \tag{88}$$

In general

$$\frac{\Delta kT}{kT_0} = -\frac{2}{5}\frac{1}{P_0}\sqrt{\frac{D_e}{\varkappa K_T}}\,\frac{u_0 - \frac{3}{2}kT - w}{\left[1 + \frac{2}{5}\frac{1-\eta}{P_0}\left(u_0 - \frac{3}{2}kT - w\right)C_s\right]^{0.5}}\Delta C_s \tag{89}$$

where $w = \left(\frac{H_a}{k} + \frac{3}{2}a\right)\frac{dkT}{da}$. Substituting the obtgined value for ΔkT into (86) we obtain for the flow $\frac{6}{d}\langle q_{ces}\rangle$ the foll wing:

$$\frac{6}{d}\langle q_{ces}\rangle = \frac{4\sqrt{3D_e}\,v^{0.5}d^{-1.5}}{\left[1 + \frac{2}{5}\frac{(1-\eta)}{P_0}\left(u_0 - \frac{3}{2}kT - w\right)C_s\right]^{0.5}}\Delta C_s \tag{90}$$

or in the form of outerdiffusional kinetic equation:

$$\frac{da}{dt} = \beta_e[C_0 - C_s^*(a)] \tag{91}$$

where the coefficient of outer mass-exchange is:

$$\beta_e = \frac{4\sqrt{3D_e}\,v^{0.5}d^{-1.5}}{\left[1 + \frac{2}{5}\frac{1-\eta}{P_0}\left(u_0 - \frac{3}{2}kT - w\right)C_s\right]^{0.5}}. \tag{92}$$

Tne value w entering into (92) is positive for the area of small values of a and negative for the area of high values f a, and at the extremum point kT it is equal to zero. Therefore, as it is evident from (92), with the increase in the value of sorption a the coefficient of outer mass-exchange β_e decreases.

The absolute value of ΔC_s and, therefore, of $\langle q_{ces} \rangle$ can decrease considerably at the first stage of the process because of the change in the temperature from its initial to its extreme value. Consequently, the velocity of the process of sorption is decreasing, as a rule, at the second stage of the process, while the first stage is of rather a short duration.

If in the equation of the outer-diffusional kinetics (91) we consider $C_s{}^*(a)$ to be the concentration in equilibrium with the surface value of sorption a_s, as it is when $T = T_0$ and not as it actually is, then for the coefficient of outer mass-exchange we will have:

$$\beta_e = \frac{4\sqrt{3D_e}\, v^{0.5} d^{-1.5}}{\left[1 + \frac{2}{5}\frac{1-\eta}{P_0}\left(u_0 - \frac{3}{2}kT - w\right)C_s\right]^{0.5}} \cdot \frac{\Delta C_s|_T}{\Delta C_s|_T} . \qquad (93)$$

The factor $\Delta C_s|_T / \Delta C_s|_{T_0}$ in (93) is equal to 1 when $t = 0$ and from then on with the rise of t rapidly decreases and can be $\ll 1$ up to $t \to \infty$ during adsorption or reverts to 1 during desorption. Therefore β_e during desorption passes minimum with an increase in a, and during adsorption β_e decreases monotonously over the whole range of changes in a. The decrease in the coefficients of outer mass-exchange β_e with the rise of a is noted in a large number of experimental studies [8, 9] and is due, apparently, to the fact that the non-isothermal process is ignored. This dependence of β_e on a is all the more defined the higher the heat of sorption and the initial concentration of the sorbate. In this case, apparently, it should be said that the process of sorption is not limited by the rate of transfer of the sorbate to the grains surface, but by the rate of withdrawal from this surface of the heat given off during the sorption.

It is characteristic that the expressions (92), (93) for the coefficient of outer mass-exchange β_e do not include in an apparent form the coefficient of thermal conductivity K_T of the non-absorbable gas. This points to the fact that independent of the value of K_T the process of sorption should always be regarded as non-isothermal.

It should be stressed that the theory of non-isothermal diffusion of the sorbate towards the grains surface through the boundary layer, employed for calculations, is principally opposed to the conclusion about the dependence of β_e on a.

The model offered here leads also to a power dependence of the mass-exchange coefficient β_e on the flow velocity v and on the grain diameter e, this dependence being confirmed by numerous experimental data [4, 10–15].

By the analysis of equation (87) and assuming the most frequent values met with in practice of the parameters included, one may disclose one more peculiarity. The point of extremum kT as a rule corresponds to the values of sorption a that differ insignificantly from the initial values. As a matter of fact, if we approximate the dependence of kT on a from the beginning of the process up to the point of extremum kT by the polynomial of the second order, then it is easy to obtain the following:

$$\left|\frac{\Delta a}{a_0}\right| \approx 2 \frac{kT_0}{u_0 - \frac{3}{2} kT_0} \left(\frac{H_a}{ka_0} + \frac{3}{2} \frac{a_m}{a_0}\right) \left|\frac{\Delta kT}{kT_0}\right|_{\max} \cdot$$

and, considering (88)

$$\left|\frac{\Delta a}{a_0}\right| \lesssim \frac{4}{5} \sqrt{\frac{D_e}{\varkappa K_T}} \left(\frac{H_a}{ka_0} + \frac{3a_m}{2\,a_0}\right) \frac{C_0}{C_{01}}.$$

The coefficient of molecular diffusion D_e and the thermal conductivity K_T are of the same order of mangitude, $\frac{H_a}{ka_0}$ is of an order of several units. Therefore, as it is stipulated that $\frac{C_0}{C_{01}} \ll 1$ then, also, $\left|\frac{\Delta a}{a_0}\right| \gg 1$. Thus, after the extremum is arrived at, the change of the temperature to the initial value falls into the considerably larger range of change in a. Then the left part of (87) for the second stage of the process isclose in order of magnitude to $\left(u_0 - \frac{3}{2} kT_0\right) \left|\frac{\Delta a}{a_0}\right|$ and may be ignored. In this case the value $w \to 0$ included into (89), (92) and (93), and the process at the second, most prolonged stage, becomes quasi-equilibrium, when practically all the heat given off or devoured during sorption is withdrawn or provided by the heat flow, respectively.

Let us now consider some quantitative laws of inner diffusion, when it puts a limit to the velocity of sorption.

As D_{mi} can be defined from (17) we may rewrite (41), (42) as:

$$\vec{q}_{ci} = - \vec{\nabla} D_i a(1 - \theta) \tag{94}$$

$$\vec{q}_{Ei} = -\vec{\nabla}(2kT - u_1) D_{mi} a(1 - \theta) - \left(\frac{H_a}{k} K_T^{(s)} + \frac{P_0}{kT} K_{Tma}\right) \vec{\nabla} kT. \tag{95}$$

At the initial moment of time $t = 0$, $\Delta kT = 0$ and

$$q_{Ei}|_{t=0} = (2kT - u_1) \frac{D_{mi}}{D_i} q_{Ci}|_{t=0}.$$

Then froj the equations for the material and heat balances (43) follows that:

$$\frac{\partial}{\partial t} \left[\frac{H_a}{k} kT + \left(\frac{3}{2} kT - \frac{D_{mi}}{D_i} 2kT - u_0 + \frac{D_{mi}}{D_1} u_1\right) a\right]\Bigg|_{t=0} = 0. \tag{96}$$

Thus, as $u_0 \gg kT$, $u_0 \gg u_1$, kT at the initial moment changes in the same direction as a. That means that during adsorption the temperature at the first stage increases. As in the inner regions of the grain the value of sorption changes more slowly than close to the surface, during the transition from the core of the grain to the periphery the temperature changes, reaches the extremum and further returns to its initial value. Hence along the radius of the grain a thermal wave forms, while in the vicinity of the radius r_m corresponding to the extremum kT, $\nabla^2 kT < 0$ during the adsorption and > 0 during desorption, which leads to the withdrawal of the heat given off during adsorption or the supply of heat devoured during desorption. Therefore in this area the temperature, after it has reached the extremum, begins to return to its initial value. Hence the thermal wave formed during sorption passes from the periphery to the core of the grains. During the motion of the wave its breadth increases and the temperature in the core of the grain changes and approaches the extremum. The sharpest peak of the temperature occurs at the initial moment of time $t = 0$ close to the grain surface and the change of the temperature has a nature of a thermal explosion. This circumstance should be taken into account when making the numerical realization for the equations of non-isothermal interdiffusional kinetics.

At the moment when the thermal wave reaches the grain core, the process of sorption passes on to its second stage, at which the distribution of temperatures in the grain is monotonous. In the core of the grain the temperature is at its maximum during adsorption and at its minimum during desorption. At this stage the temperature of all points throughout the volume of the grain changes towards its initial value. Therefore

$$|q_{Ei}| \gg \left(u_0 - \frac{3}{2}\,kT\right) |q_{ci}|$$

and, as it could be easily shown

$$|\Delta kT|_{\max} \leqq \frac{(2k_0T_0 - u_1)\dfrac{D_{mi}}{D_i} + u_0 - \dfrac{3}{2}\,kT_0}{\dfrac{H_a}{k}\,K_T{}^{(s)} + \dfrac{P_0}{kT}\,K_{Tma}}\, D_i a_0 (1 - \theta). \quad (97)$$

On the other hand, from (96) follows, approximately

$$|\Delta kT|_{\max} \approx \frac{1}{2}\,\frac{k}{H_a}\left[(2kT_0 - u_1)\,\frac{D_{mi}}{D_i} + u_0 - \frac{3}{2}\,kT_0\right] |\Delta a|. \quad (98)$$

Then, with regard to the equation of the isotherm (7) and using equations (17) and (38) for D_{mi}, D_{ma} we obtain from (97) and (98)

$$\left|\frac{\Delta a}{a_0}\right| \lesssim \frac{1}{K_T{}^{(\varepsilon)} + \dfrac{P_0}{H_aT_0}\,K_{Tma}} \left(\lambda_{mi}\sqrt{\frac{kT_0}{2\pi m}}\, e^{\frac{u_1}{kT_0}} + D_e\,\frac{B_{cma}}{B_{ce}}\right)\frac{C_0}{a_0}. \quad (99)$$

Should the heat and sorbate transfer occur only in macropores, then $\left|\dfrac{\Delta a}{a_0}\right|$ would be of the order of $\dfrac{C_0}{C_{01}}$ i.e. considerably less than 1. Actually it is rather more probable that the sorbate transfer is carried out by its diffusion inside the solid phase and the first member in the brackets in (99) is considerably larger than the second. However, in this case also, obviously, $\left|\frac{\Delta a}{a_0}\right| \ll 1$. This condition is all the more correct as the heat transfer is, as a rule, due to the heat-conductivity of the sorbent and $K_T^{(s)} \gg \frac{P_0}{H_a T_0} K_{Tma}$.

Thus the first stage of the kinetics of sorption is, as a rule, rather brief and all the more so as related to the time of sorption, the less a_0 is. Hence at the second stage the process becomes quasi-equilibrium and, approximately

$$q_{Ei} = -\left(u_0 - \frac{3}{2} kT\right) q_{ci}. \tag{100}$$

By combining (94), (95), (100) and after integrating we obtain:

$$\Delta kT = A - \frac{D_{mi}(2kT - u_1) - D_i\left(u_0 - \frac{3}{2} kT\right)}{\frac{H_a}{k} K_T^{(s)} + \frac{P_0}{kT} K_{Tma}} a(1 - \theta) \tag{101}$$

where the integration constant is

$$A = \begin{cases} \dfrac{D_{mi}(2kT_0 - u_1) + D_i\left(u_0 - \frac{3}{2} kT_0\right)}{\frac{H_a}{k} K_T^{(s)} + \frac{P_0}{kT_0} K_{Tma}} a(1 - \theta_0) & \text{during adsorption.} \\ 0 & \text{during desorption} \end{cases} \tag{102}$$

Into the expression for the coefficient of inner diffusion we introduce the value for ΔkT from (101) and obtain the dependence of D_i on the value of sorption a.

If the conditions are regular the value of sorption over the radius of the grain can be approximated by the polynomial of second order and we can approximately rewrite the equation of innerdiffusional kinetics as follows:

$$\frac{d\bar{a}}{dt} = \beta_i \frac{1}{ba_\infty} (a_s - \bar{a}) \tag{103}$$

where the coefficient of inner mass-exchange is

$$\beta_i = \frac{15}{R^2} \frac{dD_i a(1 - \theta)}{da} ba_\infty|_{a=a_s}. \tag{104}$$

And if $u_0 \gg u_1$ (101), (104) with regard to (7) we obtain

$$\beta_i = \frac{15}{R^2} \frac{\left(\frac{H_a}{k} K_T^{(s)} + \frac{P_0}{kT_0} K_{Tma}\right) D_i \, e^{\frac{u_0}{kT_0}}}{\frac{H_a}{k} K_T^{(s)} + \frac{P_0}{kT_0} K_{Tma} + \frac{\Delta u}{kT_0} \frac{u}{kT_0} \left(D_{mi} + \frac{u_0}{\Delta u} D_{ma}\right) a_s(1 - \theta_s)} . \quad (105)$$

It is obvious from (105) that β_i is in a complex way dependent on the value of sorption a_s. As a rule, with the increase in a_s the coefficient of inner mass-exchange also increases.

Hence the rate of desorption is usually considerably less than the rate of adsorption. As with the rise in a_s β_e decreases and β_i increases then if the values of $\bar{a}$ are small the process of sorption proceeds in the outerdiffusional region and if the values of $\bar{a}$ are high in the interdiffusional regions.

4. Conclusion

The anthors presented a model of sorption over active centers as potential boxes and also considered the mechanism of heat-mass-transfer due to the wandering of molecules over the active centers.

On the basis of BOLTZMANN's kinetic equation we obtained the equation for heat and mass transfer in macropores and close to the grains' surface.

It is shown that the assumption of the isothermal natural of sorption pro cesses is fundamentally opposed to the model employed and the process, therefore, should always be regarded as nonisothermal.

There are also shown two stages in the developement of the process of sorption. At the first relatively brief stage for the outer-diffusional conditions there occurs a rapid rise in temperature up to its extreme value. For the innerdiffusional conditions inside the grain there arises a thermal wave that moves from the surface of the grain towards its center. At the second, quasi-equilibrium stage such a correlation between the heat and mass flows starts that opposes a rapid increase or decrease of heat in any element of volume of the system. At this stage the temperature from its extremum monotonously speeds towards its initial value. There are also obtained approximate estimates for the extreme temperatures.

For the second and more lengthy stage two expressions are obtained for the coefficient of outer and inner-mass-transfer, and it should be noted that with the rise in the value of sorption β_e decreases, while β_i increases.

5. Basic designations

λ_{tur}, t_{rel} — turbulency scale and relaxation time
t — time
v — carrier-flow velocity; velocity of the thermal movement of molecules

T, T_s, T_0, $\bar{T}$	– temperature of current, surface, initial and mean over the grain volume,
R	– BOLTZMANN's constant
S	– entropy
a, a_∞, a_s, a_v^*, $\bar{a}$	– value of sorption, current, limit, surface, characteristic and mean
C, C_s, C_0, $C^*(a)$	– concentration of the sorbate, current, surface, characteristic and equilibrium
C_1	– concentration of molecules of the carrier
b	– isotherm parameter
u_0, Δu, u_1	– exit work of the absorbed molecule, height of the potential barrier between the neighbouring active centers and a difference between the exit work and the height of the potential barrier.
λ_{mi}, λ_{ma}	– distance between the neighbouring active centers and length of free-path for the molecules of rarefied gas in the macropores
S_{ma}, V_{ma}	– specific surface and volume of the macropores
t_a	– mean time of the uninterrupted stay of a molecule on an active center
m, m_1	– mass of the molecules of the sorbate and the carrier-gas
V	– mean velocity of movement of the molecules inside the gaseous phase
f, f_0, f_1, f_{10}	– function of distribution of the molecules by their velocities and its zero approximation for the sorbate and the carrier
σ_{01}, σ_{11}	– section of collisions between the molecules of the carrier with the molecules of the sorbate and with each other
ν_{ma}, ν_{01}, ν_{11}	– frequency of collisions of molecules with the macropores surface and of isolated molecules of the sorbate and the carrier with molecules of the carrier
q_{cmi}, q_{cma}, q_{ci}, q_{ce}, $q_{ce}^{(g)}$	– sorbate flows in the solid phase, macropores, total flow inside a grain and in the gaseous phase, and the molecules of the carrier flow
q_{mv}, q_{mvma}	– impulse flows in the carrier and in the macropores
D_{mi}, D_{ma}, D_i, D_e	– coefficients of molecular diffusion of the sorbate in the solid phase, macropores, total inside the grain and in the carrier
$K_T^{(s)}$, K_{Tma}, K_T	– coefficients of thermal conductivity of the sorbent and carrier-gas in the macropores and outside a grain
H_a	– thermal capacity of the sorbent
P	– pressure of the carrier-gas
d, R	– diameter and radius of the sorbent grain
β_e, β_i	– coefficient of outer and inner mass-exchange.

6. References

[1] BISCHOFF, K. B., LEVENSPIEL, O., Chem. Eng. Sci., **17**, 245, 257, (1962)

[2] CARBERRY, I. I., WILHELM, R. H., Am. Inst. Chem. Eng. I., **4**, 367 (1958)

[3] LEVENSPIEL, O., "*Chemical Reaction Engineering*", Wiley, New York 1962

[4] AEROV, M. E., TODES, O. M., "*Hydrodynamic and Thermal Foundation of Fixed and Fluidized Bed Apparatous*", Chemistry, Moscow 1968

[5] BOLTZMANN, L., „*Vorlesungen über Gastheorie*", Leipzig 1896

[6] CHAPMAN, S., COWLING, T. G., "*The Mathematical Theory of Non-Uniform Gases*", 2nd ed, Cambridge, 1952

[7] HUANG, K., *"Statistical Mechanisc"*, New York—London 1963
[8] DUBININ, M. M., LESIN, JU. S., DAN USSR **171**, 382 (1966)
[9] ZABEJINSKI, Ja. L., JUCHOVITSKI, A. A., TICHONOV, A. N., J. Phys. Chem. **23**, 92 (1949)
[10] KUTADELADSE, S. S., *"Theoretical Foundation of Heat Exchange"*, Mashgis, Moscow 1957
[11] KUSIK, C. L., HAPPEL, I., Ind. Chem. Fund. **1**, 163 (1962)
[12] CARBERRY, I. I., Am. Inst. Chem. Eng. J. **6**, 460 (1960)
[13] SHUCHOVITZKI, A. A., ZABESHINZKI, JA. L., VENISHKINA, A. I., J. Phys. Chem., **15**, 174 (1941)
[14] ZABESHINSKI, Ja. L., J. Phys. Chem. **17**, 32 (1943)
[15] LINTON, M., SUTHERLAND, K. L., Chem. Eng. Soc. **12**, 214 (1960)

On sorption dynamic theory for great concentrations of adsorbed component

(V. L. Kolin, E. A. Ustinov, I. N. Taganov)

Zusammenfassung:

Es werden mathematische Modelle isothermer Sorptionsprozesse im Festbett und in dichten Wanderschichten mit einer von der momentanen Sorptivkonzentration abhängigen Gasstromgeschwindigkeit vorgeschlagen. Die Besonderheit dieser Prozesse wird untersucht und festgestellt, daß die mit dem Ansteigen des Adsorptionsgrades verbundene Geschwindigkeitsverringerung zu einer Kompression der Sorptionsfront führt. Numerische Lösungsmethoden und durchgerechnete Beispiele werden angeführt. Für das asymptotische Verhalten und andere Grenzsituationen wurden analytische Lösungen erhalten, die verfahrenstechnische Berechnungen zulassen. Die Berechnungsergebnisse werden mit experimentellen Daten zur Adsorption von Normalparaffinen aus Kerosinfraktionen des Erdöls verglichen.

General concepts of adsorption dynamics of a gas flow are developed for the case when the final product is contained in the flow-carrier in a minute amount and the change of its concentration does not result in the change of flow density or mass and volume [1]. This restriction is not essential for many adsorption processes. However, there is a series of processes in which the initial content of the final product in the mixture comes up to a few decades of volumetric percent and if the required extent of recovery approaches 100% the dependence of flow rate upon the final product current cannot be neglected. Some general regularities of such processes were considered in the reports [2, 3]; however, methods of engineering estimation with allowance for kinetic regularities have not been developed yet.

Let us consider an isothermic adsorption process taking place under constant total pressure when the relation of sorbing component concentration and sorbing gas in any section of the layer is estimated with the help of Mendeleev-Clapeyron's law.

$$C + C' = \frac{P_0}{RT}. \tag{1}$$

Where C, C' — mole concentrations of sorbing component and gas, respectively, P_0 — pressure in the system; R — gas constant; T — temperature, °K. Let us suppose that the gas is sorbed in the layer in negligibly minute amount and the process is accomplished with the help of molecular-screen effect. When the influence of longitudinal effects is insignificant material balance equations for adsorbing and non-adsorbing components of the mixture can be written in the following way:

$$(1 - \varepsilon) \frac{\partial a}{\partial t} + \varepsilon \frac{\partial C}{\partial t} + \frac{\partial vC}{\partial x} = 0, \tag{2}$$

$$\varepsilon \frac{\partial C'}{\partial t} + \frac{\partial vC}{\partial x} = 0. \tag{3}$$

Where:

a — adsorption value related to the unit of sorbent volume;
v — linear flow velocity designed for the total layer section;
x, t — current values of the coordinate and time, respectively;
ε — layer porosity.

Adding equations (2) and (3) and taking into consideration (1) we obtain

$$(1 - \varepsilon) \frac{\partial a}{\partial t} + \frac{P_0}{RT} \frac{\partial v}{\partial x} = 0. \tag{4}$$

If the relation of the adsorption value to the corresponding sorbate equilibpium concentration in the flow is sufficiently high it can be shown that $\frac{\partial c}{\partial t} \ll \frac{\partial a}{\partial t}$. Having excluded $\frac{\partial a}{\partial t}$ from equations (2), (4) and taking into consideration that $x = 0$, $C = C_0$, $C' = C_0'$, $v = v_0$ we obtain after integration the dependence of the flow rate on the current sorbate concentration

$$v = v_0 \frac{C_0'}{C_0 + C_0' - C}. \tag{5}$$

If the adsorption equilibrium is established so readily that the process can be regarded as an equilibrium one, current values of a and C are related to each other by an isotherm equation. In this case the equation for the speed of the movement of the sorption front points in which $C =$ const. results from (2)

$$u = \frac{v^2(C_0 + C')}{(1 - \varepsilon) v_0 C' a'(C)}, \tag{6}$$

where $a'(C)$ — derivative of the adsorption isotherm. Having substituted v from (5) into (6) we obtain

$$u = \frac{v_0 C_0'(C_0 + C_0')}{(C_0 + C_0' - C)^2 (1 - \varepsilon) a'(C)} \tag{7}$$

which permits to investigate the movement of the concentration front points over the layer depending on the isotherm curvature and the direction of C change. In a particular case $v = v_0 =$ const. a well-known WICKE's law [4] results from (7).

When the isotherm is linear and $a'(C) = \Gamma =$ const u increases with C. Consequently, the concentration front is contracted by the desorption. When the isotherm is convex and $a'(C_1) < a'(C_2)$ for $C_1 > C_2$ the change of $v = v(C)$

is an additional factor contributing to concentration and its wash-out under desorption. When $C_1 > C_2$, $a'(C)_1 > a'(C_2)$, i.e. the isotherm is concave, the change of u depending on isotherm curvature and relation $\frac{C_0}{C_0'}$ can result in low front contraction under adsorption and front wash-out under desorption.

For example, when the sorption isotherm is defined by the LANGMUIR equation

$$u = \frac{v_0 C_0'(C_0 + C_0')(1 + bC)^2}{(1 - \varepsilon)(C_0 + C_0' - C)\, a_\infty b} \tag{8}$$

the speed of movement of the front points $C \to C_0$ and $C \to 0$ is equal to

$$\frac{u_{C \to C_0}}{u_{C \to 0}} = (1 + bC_b)^2 \left(1 + \frac{C_0}{C_0'}\right)^2 = \left[\frac{(1 + p)(1 + \lambda)}{p}\right]^2 \tag{9}$$

where $p = 1/bC_0$ — parameter characterising isotherm curvature, $\lambda = \frac{O_0}{C_0'}$. For the isotherms which are convex towards concentration axis this parameter is $p \in (0, \infty)$, for concave isotherms — $p \in (-1, -\infty)$ and for linear ones $-p \to \pm\infty$.

The condition providing the possibility of concentration front formation where $\frac{u_{C \to C_0}}{u_{C \to 0}} \simeq 1$ results from (9), i.e. all points move with constant speed. This condition can be written in such a way:

$$\left|\frac{1 + p}{p}(1 + \lambda)\right| \gg 1. \tag{10}$$

Thus, in contrast to adsorption processes with low initial concentration of the component to be sorbed in the flow the considered formation of quasi-stationary concentration front moving with the constant speed could occur in the case of convex, linear and even concave isotherms.

In many cases the application of asymptotic dependences to technological adsorption processes estimation is auite reasonable. As the flow rate decreases with the increase of the final component the recovery rate is a factor contributing to adsorption front contraction and consequently to the diminishing of zone length and time of quasistationary regime formation it can be assumed that asymptotic dependences of the process will be the most essential factors for the solution of engineering problems.

As the choice of an equation describing adsorption kinetics depends on concrete system properties and conditions of process accomplishment it is reasonable to consider the dependences of unbalanced adsorption applicable to the various equations of kinetics. Let us confine ourselves to the consideration of the adsorption isotherm described by the LANGMUIR equation and by the following equations of kinetics.

$$\frac{\partial a}{\partial t} = \beta_1[C - C^*(a)], \tag{11}$$

$$\frac{\partial a}{\partial t} = \beta_2[a^*(C) - a], \tag{12}$$

$$\frac{\partial a}{\partial t} = \beta_3 C\left(1 - \frac{a}{a_\infty}\right) - \frac{\beta_3}{b}\frac{a}{a_\infty}. \tag{13}$$

Then systems of corresponding equations describing front process dynamics can be represented in the following way

$$\frac{\partial \gamma}{\partial \tau} + \frac{1}{\Gamma}\frac{\varepsilon}{1-\varepsilon}\frac{\partial u}{\partial \tau} + \frac{\partial \alpha y}{\partial \xi} = 0, \tag{14}$$

$$\frac{\partial \gamma}{\partial \tau} + \frac{1+\lambda}{\lambda}\frac{\partial \alpha}{\partial \xi} = 0, \tag{15}$$

$$\frac{\partial \gamma}{\partial \tau} = y - \frac{p\gamma}{1 + p - \gamma}. \tag{16}$$

Equations (17) or (18) can be used instead of (16)

$$\frac{\partial \gamma}{\partial \tau} = y\frac{p}{p+y} - \frac{p}{p+1}\gamma, \tag{17}$$

$$\frac{\partial \gamma}{\partial \tau} = y\left(1 - \frac{\gamma}{1+p}\right) - \frac{p}{p+1}\gamma. \tag{18}$$

Initial and boundary conditions of the front process

$$\xi \geqq 0, \quad \tau = 0, \quad \gamma = 0; \quad \xi = 0, \quad \tau \geqq 0, \quad y = 1.$$

Where:

$$\xi = \frac{\beta_i}{v_0}(1-\varepsilon)\,x; \quad \tau = \frac{\beta_i}{\Gamma}t; \quad y = \frac{C}{O_0}; \; \gamma = \frac{a}{a_0}; \quad \alpha = \frac{v}{v_0};$$

$$\Gamma = \frac{a_0}{C_0}; \quad i = 1, 2, 3;$$

β_i — kinetic coefficient;
b — LANGMUIR equation constant;
a_∞ — boundary adsorption value;
C_0, C_0' — initial concentration of adsorbina component and gas-carrier, respectively;
a_0 — adsorption value equilibrium with C_0;
v — linear speed of flow at the input to the layer.

As the front moving with the speed $u = \text{const}$ is formed in the layer under sufficiently large space-time removal from the beginning of the process, it is possible to introduce a new variable $z = u\tau - \xi$. Then instead of (12), (14),

(15) we obtain

$$u \frac{d\gamma}{dz} + \frac{u}{\Gamma} \frac{\varepsilon}{1-\varepsilon} \frac{dy}{dz} - \frac{d\alpha y}{dz} = 0, \tag{19}$$

$$u \frac{d\gamma}{dz} - \frac{1+\lambda}{\lambda} \frac{d\alpha}{dz} = 0, \tag{20}$$

Integration of (49), (20) gives:

$$u\gamma + \frac{u}{\Gamma} \frac{\varepsilon}{1-\varepsilon} y - \alpha y = A, \tag{21}$$

$$u\gamma - \frac{1-\lambda}{\lambda} \alpha = B, \tag{22}$$

where A, B — integration constants.

As with $z \to -\infty$ $(\xi \gg u\tau)$ $y \simeq \gamma \simeq 0$, $A = 0$. When $z \to \infty$ the sorbent layer is purified to complete saturation and $y \simeq \gamma \simeq 1$, $\alpha = 1$ and consequently the relative sorption front speed of movement is

$$u = \frac{(1-\varepsilon)\Gamma}{(1-\varepsilon)\Gamma + \varepsilon}. \tag{23}$$

Hence the quasistationary sorption front speed of movement does not depend on the low change of flow linear speed in the layer and when $\Gamma > 1$ $u \simeq 1$.

Having substituted u from (23) into (22) and considering that when $z \to \infty$ $\gamma \simeq \alpha \simeq 1$ we obtain

$$B = \frac{(1-\varepsilon)\Gamma}{(1-\varepsilon)\Gamma + \varepsilon} - \frac{1+\lambda}{\lambda}.$$

Thus instead of (21), (22) we have

$$\frac{(1-\varepsilon)\Gamma}{(1-\varepsilon)\Gamma + \varepsilon} \gamma + \frac{\varepsilon}{(1-\varepsilon)\Gamma + \varepsilon} y - \alpha y = 0 \tag{24}$$

$$\frac{(1-\varepsilon)\Gamma}{(1-\varepsilon)\Gamma + \varepsilon} \gamma - \frac{1+\lambda}{\lambda} \alpha = \frac{(1-\varepsilon)\Gamma}{(1-\varepsilon)\Gamma + \varepsilon} - \frac{1+\lambda}{\lambda}. \tag{25}$$

Having excluded α from (24), (25) we obtain the relation of the absorbed component concentrations in the adsorbent and flow-carrier within the sorption front

$$\gamma = \frac{y}{1 + \lambda - \lambda y}. \tag{26}$$

When $\lambda \to 0$, i.e. when the adsorption occurs in the flows with low content of adsorbed component a well-known relationship of ZELDOVICH [1]: $y \simeq \gamma$ results from (26).

Now let us consider concentration distribution within the quasistationary sorption front with the kinetics defined in (16), (17), (18).

Having accomplished conversion to independent variable $z = u\tau - \xi$ in equations (16)—(18) and substituted y from (26) into each of the equations of kinetics and integrated them we obtain

$$\ln \frac{\gamma}{(1+\gamma)\dfrac{(1+p)(1+\lambda)}{1+p}} + \frac{\lambda\gamma^{\circ}}{1+p} = \frac{1+\lambda+p\lambda}{1+p}\left(1+\frac{1}{\Gamma}\frac{\varepsilon}{1-\varepsilon}\right)z + A_1, \tag{27}$$

$$\ln \frac{\gamma}{(1-\gamma)\dfrac{(1+p)(1+\lambda)}{p}} = \frac{1+\lambda+p\lambda}{1+p}\left(1+\frac{1}{\Gamma}\frac{\varepsilon}{1-\varepsilon}\right)z + A_2, \tag{28}$$

$$\ln \frac{\gamma}{(1-\gamma)^{1+\lambda}} = \frac{1+\lambda+\lambda p}{1+p}\left(1+\frac{1}{\Gamma}\frac{\varepsilon}{1-\varepsilon}\right)z + A_3, \tag{29}$$

where A_1, A_2, A_3 — integration constants. In order to determine integration constants let us use the material balance equation of the adsorbed component, which for the quasistationary condition can be written in such a way

$$\int_0^{-\infty}\left(\gamma+\frac{1}{\Gamma}\frac{\varepsilon}{1-\varepsilon}y\right)dz = \int_{\infty}^{0}\left(1+\frac{\gamma}{\Gamma}\frac{\varepsilon}{1-\varepsilon}-\gamma-\frac{y}{\Gamma}\frac{\varepsilon}{1-\varepsilon}\right)dz. \tag{30}$$

Hence expression for integration constants is obtained

$$A_1 = \frac{\Gamma\lambda(p+0.5)(1-\varepsilon)}{(p+1)[(1-\varepsilon)\Gamma+\varepsilon]}, \tag{31}$$

$$A_2 = \frac{\Gamma(1-\varepsilon)}{\Gamma(1-\varepsilon)+\varepsilon}\frac{1+\lambda+p\lambda}{p} + \frac{\Gamma(1-\varepsilon)}{\Gamma(1-\varepsilon)+\varepsilon}\frac{1+\lambda}{p\lambda}\ln(1+\lambda), \tag{32}$$

$$A_3 = \frac{\lambda\Gamma(1-\varepsilon)}{\Gamma(1-\varepsilon)+\varepsilon}. \tag{33}$$

Thus on the quasistationary stage of process expressions similar to Shilov's equation [5] are valid.

$$\tau = \left(1+\frac{1}{\Gamma}\frac{\varepsilon}{1-\varepsilon}\right)\xi - \tau_0, \tag{34}$$

where the dimensionless lag of "safety action time" τ_0 is

$$\tau_0 = \frac{\lambda(p+0.5)(1-\varepsilon)\Gamma}{(1+\lambda+p\lambda)[(1-\varepsilon)\Gamma+\varepsilon]} - \frac{1}{1+\lambda+p\lambda} \times \left[1+\lambda\gamma-(1+p)\ln\frac{(1-\gamma)^{\frac{(1+\lambda)p}{1+p}}}{\gamma}\right]$$

with the kinetics according to (16),

$$\tau_0 = \frac{1+p}{p} \frac{\Gamma(1-\varepsilon)}{\Gamma(1-\varepsilon)+\varepsilon} \left[1 + \frac{(1+\lambda)\,\varepsilon}{(1-\varepsilon)\,\Gamma\lambda(1+\lambda+p\lambda)} \ln(1+\lambda)\right]$$

$$+ \frac{1+p}{1+\lambda+p\lambda} \ln \frac{(1-\gamma)^{\frac{(p+1)(1+\lambda)}{p}}}{\gamma}$$

with the kinetics according to (17),

$$\tau_0 = \frac{1+p}{1+\lambda+p\lambda} \left[\frac{\Gamma\lambda(1-\varepsilon)}{\Gamma(1-\varepsilon)+\varepsilon} \ln \frac{(1-\gamma)^{1+\lambda}}{\gamma}\right]$$

with the kinetics of "LANGMUIR type" (18). In a particular case of linear isotherm ($p \to \pm\infty$) expressions for τ_0 become identical and

$$\tau_0 = \frac{1}{\lambda} \left[\frac{\lambda\Gamma(1-\varepsilon)}{\Gamma(1-\varepsilon)+\varepsilon} + \ln \frac{(1-\gamma)^{1+\lambda}}{\gamma}\right].$$

Naturally, limiting conversion to $\lambda \to 0$ does not occur in this case.

Relationships obtained for the asymptotic stage of the process can be applied in the engineering estimations.

Conclusion

1. In the present paper the problem of the dynamics of component adsorption from the mixture under constant pressure and flow-rate depending on current component concentration is studied. Possibility and conditions of quasistationary front formation with convex, linear and concave isotherms are considered.

2. Quasistationary sorption front speed of movement does not depend upon the relationship of concentrations in the flow and is estimated by the expression applicable for adsorption in a flow with low concentration of sorbing component. However, concentration relationship of the sorbed substance in the flow and the flow and the adsorbate within the front asymptotically agrees with the relationship of ZELDOVICH.

3. General dependences of asymptotic state, permitting to determine the operting time of the layer on the defined concentration of the adsorbed component, mass-exchange zone length and other dynamic characteristics of the process were obtained for various equations of sorption kinetics.

References

[1] Rachinsky, V. V., "*Introduction into General Theory of Sorption and Chromatography Dynamics*", Nauka, Moscow 1964
[2] Schay, G., „*Theoretische Grundlagen der Gaschromatographie*", VEB Deutscher Verlag der Wissenschaften, Berlin 1961 (Hrsg.: E. Krell)
[3] Schay, G., Collection of works "*Development of Chromatography*", p. 179, Nauka, Moscow 1972
[4] Wicke, E., Koll. Z., 86, No. 1, 67, 295 (1939)
[5] Shilov, A. N., Lepin, L. K., Voznesensky, S. A., J. Russ. Phys. Chem. Soc. 61, 1107 (1929)

A study of unlinear chromatography

(E. A. Ustinov, V. L. Kolin, I. N. Taganov, A. A. Seballo, N. S. Frolova)

Zusammenfassung:

Der Beitrag beschäftigt sich mit dem isothermen irreversiblen Sorptionsmodell mit Längsvermischungseffekten bei Isothermen vom Langmuir-Typ.

Es werden die Gesetzmäßigkeiten der Verdrängungs- und Elutionschromatographie untersucht. Für das asymptotische Verhalten wurden analytische Lösungen für das Gleichungssystem bei äußerer, innerer und Übergangskinetik gefunden. Es wurden die Abhängigkeiten der Form der Peaks und die Zeitpunkte ihres Auftretens von den Parametern der Isotherme und anderen Prozeßbedingungen ermittelt. Für die Chromatographie von Gemischen werden die numerisch gewonnenen Lösungen vom Standpunkt der reversiblen und irreversiblen Modellvorstellungen analysiert.

1. Introduction

The laws of the development of chromatographic processes have been in the main part determined for the linear isotherm of sorption. If the isotherm is unlinear then it is only possible as a rule to employ numerical methods because analytical solutions of the problems of the dynamics of chromatographical processes can be obtained only for some special cases. However, it is obvious that the analytical form of the system of differential equations for the dynamics of a developing or eluent process is much more preferable than a numerical solution. Though the analytical solution does not increase the informative content of the employed mathematical model, all the same it makes it possible to define in the most obvious form the correlations between various parameters and quantities being measured. Also, and this is vastly important for chromatographic processes, the analytical solutions allow one to determine the asymptotic laws.

2. Analysis of Chromatographic Processes, based on the Analytical Solutions of the Problems of Dynamics

Let us examine a case of the influence of longitudinal effect being small and the process of sorption described by the following system of the hyperbolic type:

$$\frac{\partial a}{\partial t} = \beta_3 C \left(1 - \frac{a}{a_\infty}\right) - \frac{\beta_3}{b} \frac{a}{a_\infty} \tag{1}$$

$$\frac{\partial a}{\partial t} + \frac{\partial C}{\partial t} + v \frac{\partial C}{\partial x} = 0. \tag{2}$$

And in a dimensionless form, instead of (1) and (2)

$$\frac{\partial \gamma}{\partial \tau} = y\left(1 - \frac{1}{p+1}\gamma\right) - \frac{p}{p+1}\gamma \tag{3}$$

$$\frac{\partial y}{\partial \xi} = \frac{p}{p+1}\gamma - y\left(1 - \frac{1}{p+1}\gamma\right). \tag{4}$$

The initial and boundary conditions, which allow for the correct stating of the problem by Adamard are as follows:

$$\begin{aligned} &\tau = 0, \quad \xi \geqq 0, \quad \gamma = \gamma(\xi, 0) \\ &\tau \geqq 0, \quad \xi = 0 \end{aligned} \tag{5}$$

where

$$\gamma = \frac{a}{a_0}; \quad y = \frac{C}{C_0}; \; \xi = \frac{\beta_э x}{v}; \quad \tau = \frac{\beta_э C_0}{a_0}\left(t - \frac{x}{v}\right); \quad p = \frac{1}{bC_0};$$

C_0 being the characteristic concentrations in the flow and a_0 being the value of sorption in equilibrium.

We linearize the system (3), (4) by the substitution:

$$\gamma = \gamma(\xi, 0) - (p+1)\frac{1}{\varrho}\frac{\partial \varrho}{\partial \xi}; \quad y = (p+1)\frac{1}{\varrho}\frac{\partial \varrho}{\partial \tau}. \tag{6}$$

Then we may re-write (3)—(5) as to ϱ in the following form:

$$\frac{\partial^2 \varrho}{\partial \xi\, \partial \tau} + \frac{\partial \varrho}{\partial \tau}\left(1 - \frac{1}{p+1}\gamma(\xi, 0)\right) + \frac{p}{p+1}\frac{\partial \varrho}{\partial \xi} - \frac{p}{p+1}\gamma(\xi, 0)\,\varrho = 0 \tag{7}$$

$$\tau = 0, \quad \xi \geqq 0, \quad \varrho = 1$$

$$\tau \geqq 0, \quad \xi = 0 \quad \varrho = \exp\frac{1}{p+1}\int_0^\tau y(0, \tau)\, d\tau. \tag{8}$$

Let us take a case of development of chromatography if the introduction of the sample in the column is performed step by step during the period of time τ_0 and when the initial content of the sorbate in the bed is nil, that is

$$\gamma(\xi, 0) = 0; \quad y(0, \tau) = \begin{cases} 1, & \text{if } 0 \leqq \tau \leqq \tau_0 \\ 0, & \text{if } \tau_0 < \tau < \infty, \end{cases} \tag{9}$$

Taking (9) into account we get for y, γ from (7)

$$y = \frac{y_l\left(\frac{p}{p+1}\,\xi, \tau\right) - y_l\left(\frac{p}{p+1}\,\xi, \tau - \tau_0\right)}{y_l\left(\frac{p}{p+1}\,\xi, \tau\right) - y_l\left(\frac{p}{p+1}\,\xi, \tau - \tau_0\right) + e^{\frac{\xi-\tau}{p+1}}\left[1 - y_l\left(\xi, \frac{p}{p+1}\,\tau\right) + e^{\frac{\tau_0}{p+1}}\, y_l\left(\xi, \frac{p}{p+1}\,(\tau - \tau_0)\right)\right]} \tag{10}$$

$$\gamma = \frac{\gamma_l\left(\frac{p}{p+1}\,\xi, \tau\right) - \gamma_l\left(\frac{p}{p+1}\,\xi, \tau - \tau_0\right)}{y_l\left(\frac{p}{p+1}\,\xi, \tau\right) - y_l\left(\frac{p}{p+1}\,\xi, \tau - \tau_0\right)}\, y. \tag{11}$$

The functions $y_l(\xi, \tau)$, $\gamma_l(\xi, \tau)$ in the equations (10), (11) are the solutions of the problem of the dynamics of sorption for the frontal conditions when the isotherm is linear, that is

$$y_l(\xi, \tau) = e^{-(\xi+\tau)} \sum_{n=0}^{\infty} \frac{\xi^n}{n!} \sum_{m=n}^{\infty} \frac{\tau^m}{m!}, \qquad \gamma_l(\xi, \tau) = e^{-(\xi+\tau)} \sum_{n=0}^{\infty} \frac{\xi^n}{n!} \sum_{m=n+1}^{\infty} \frac{\tau^m}{m!}. \tag{12}$$

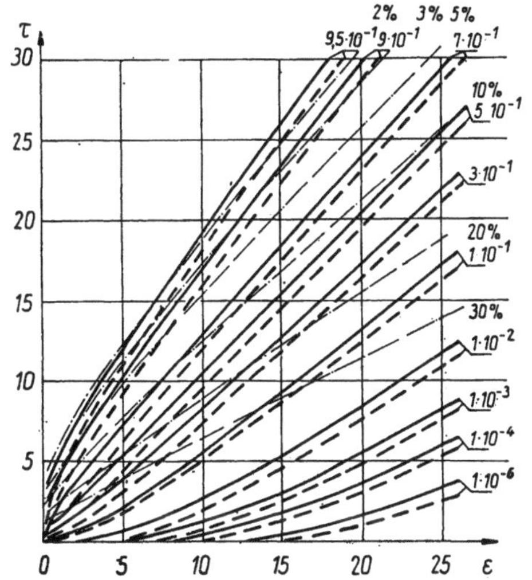

Fig. 1: Dynamical curves for frontal sorption process with the linear isotherm; $p \to \infty$.

- - - y; ——— γ; —·— $\frac{y-\gamma}{\gamma}$

These functions are shown graphically in Fig. 1. The equations (12) can be also employed for the analysis of chromatographic processes with linear isotherm.

For example, in the case of development chromatography

$$y(\xi, \tau) = y_l(\xi, \tau) - y_l(\xi, \tau - \tau_0)$$
$$\gamma(\xi, \tau) = \gamma_l(\xi, \tau) - \gamma_l(\xi, \tau - \tau_0) \tag{13}$$

and for eluent chromatography if we have the step-by-step law of the initial distribution of substance over the layer and the initial concentration is zero, then:

$$\gamma(\xi, 0) = \begin{cases} 1, & \xi \leqq \xi_0, \\ 0, & \xi > \xi_0, \end{cases} \quad y(0, \tau) = 0$$
$$y(\xi, \tau) = y_l(\xi - \xi_0, \tau) - y_l(\xi, \tau)$$
$$\gamma(\xi, \tau) = \gamma_l(\xi - \xi_0, \tau) - \gamma_l(\xi, \tau). \tag{14}$$

In a similar fashion we can construct the superpositions of the functions (12) for other more complicated cases. An example of how the function (12) can be employed to present chromatographical processes having linear isotherm is shown in Fig. 2.

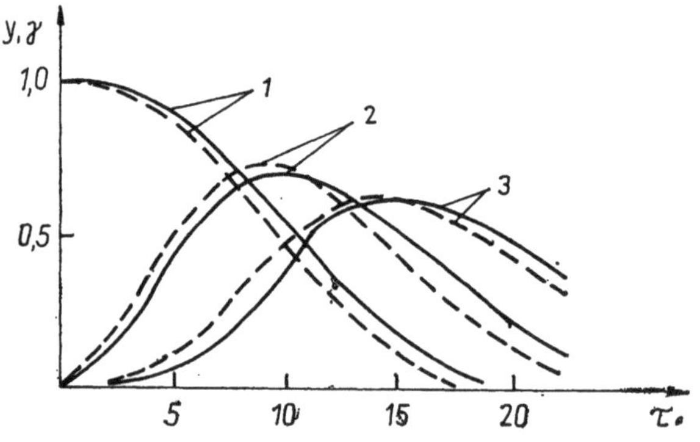

Fig. 2: Evolution of the eluent chromotographical process. Calculation for data of Fig. 1.
$\xi_0 = 10$; ξ: 1 — 10; 2 — 15 3 — 20.

The expressions (10), (11) also allow a close view on rules by which the chromatographic stripe is formed for both convex and concave isotherms. Fig. 3, 4 give as an example results of calculations which show the development of the process when the isotherm is convex and also show the influence of the sample volume.

At the asymptotic stage of the process when $x \to \infty$, $t \to \infty$, the expressions (10), (11) become indefinite. Using the properties of the functions contained in (10), (11) after some simple transpositions we can obtain the following

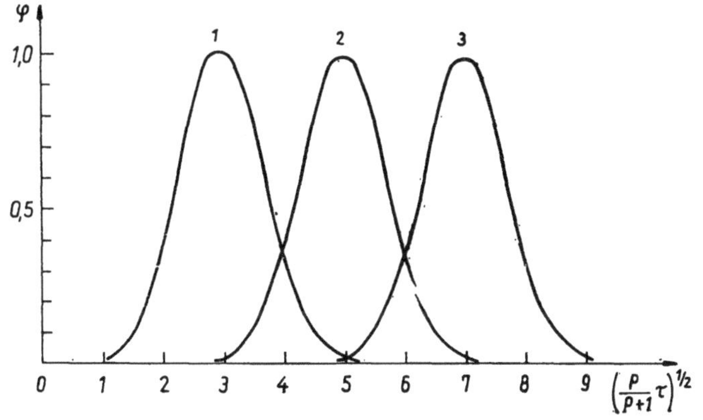

Fig. 3: Evolution of the developing chromotographical process for different lengths of the column.
$\sqrt{\xi}$: 1 – 3; 2 – 5; 3 – 7; $\alpha = 1$.

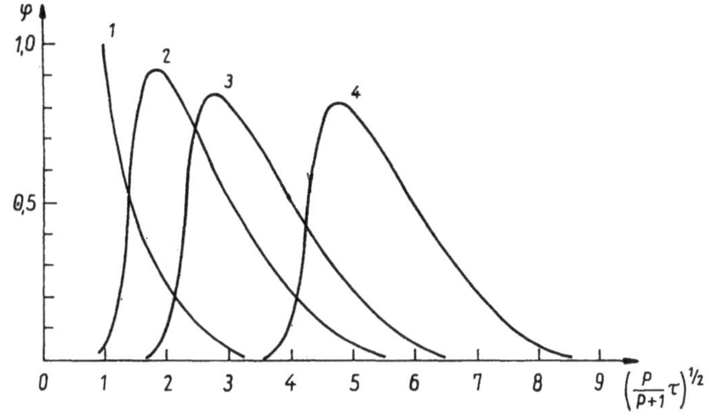

Fig. 4: Dynamical curves for developing chromotographical process.
$\sqrt{\xi}$: 1 – 2; 2 – 4; 3 – 5; 4 – 7. $\alpha = 10$.

solution for the asymptotic condition:

$$y \cong \sqrt{\frac{p\xi}{(p+1)\,\tau}}\,\gamma \cong \frac{1}{2}\sqrt{\frac{p(p+1)}{\pi\tau}}\;\frac{e^{-\left(\sqrt{\xi}-\sqrt{\frac{p\tau}{p+1}}\right)^2}}{\dfrac{1}{e^{\frac{\tau_0}{p+1}}-1}+\dfrac{1}{2}\operatorname{erfc}\left(\sqrt{\xi}-\sqrt{\dfrac{p\tau}{p+1}}\right)}. \tag{15}$$

Equation (15) shows that the stripe fades with time in proportion to $\sqrt{t}$. The widths of the stripe, on the contrary, lessen in reverse proportion to $\sqrt{t}$.

It is convenient to have equation (15) in the following form

$$\varphi = \frac{1}{\alpha} \cdot \frac{e^{-z^2}}{\frac{1}{e^\alpha - 1} + \frac{1}{2}\operatorname{erfc}(z)}, \tag{16}$$

where

$$z = \sqrt{\xi} - \sqrt{\frac{p}{p+1}\tau}; \quad \varphi = \frac{2}{\alpha p}\sqrt{\pi \frac{p}{p+1}\tau}\, y; \quad \alpha = \frac{\tau_0}{p+1} = \frac{\beta_э Q}{vsa_\infty}$$

Q — the volume of the sample introduced into the column;
S — the column's cross-section.

The form of the curve φ in the $\varphi - z$ coordinates as is shown in (16) depends for the asymptotic stage only on α. The stripe is in general asymmetry and the more so the higher the value of α. The asymmetry increases with the increase in the curvature of the isotherm, with the increase in the volume of the sample, introduced into the column, with the rise in the mass-exchange coefficient and the lessening of the carrier-gas consumption. If the isotherm is concave the concentration in the front of the stripe changes more drastically than in the tail-part. The stripe moves at the rate of v/ba_∞ but the maximum rate of movement occurs, contrary to the case of the linear isotherm, sooner after the layer than may be expected on the condition $x = vt_m/ba_\infty$. This is due to the fact that at the stage where the stripes are formed and the concentrations of the component in the flow are sufficiently large, the rates of movement of the concentration points are higher than at the asymptotic stage because the mass-exchange is less intense at this stage. At the asymptotic stage of the process the rates or the concentration points are equal in value and constant in time, but with the drift to the area of lesser and lesser concentrations the stripe begins to fade in proportion to $\sqrt{t}$ and its peak is lessening accordingly. Hence, when x and t are large the form of the stripes remains constant in the $C\sqrt{t} - \sqrt{t}$ coordinates and the stripes only shift towards the increase in $\sqrt{t}$ with the increase in the length of the column. For the isotherm in question such behaviour of the stripes is preserved also when there are longitudinal effects and also when there is a different equation for the kinetics of sorption. Suppose the process takes place in a mixed-diffusional area, then the system of equations would be:

$$\frac{\partial a}{\partial t} = \beta_e \left[C - \frac{1}{b} \frac{a_s}{a_\infty - a_s} \right] \tag{17}$$

$$\frac{\partial a_i r}{\partial t} = D_i \frac{\partial^2 a_i r}{\partial r^2}, \quad a = \frac{3}{R^2} \int_0^R a_i r^2 dr \tag{18}$$

$$\frac{\partial a}{\partial t} + \frac{\partial C}{\partial t} + v \frac{\partial C}{\partial x} = D_э \frac{\partial^2 C}{\partial x^2}. \tag{19}$$

For the asymptotic stage it follows (approximately) from (18)

$$\frac{\partial a}{\partial t} = \beta_i \frac{1}{ba_\infty}(a_s - a), \quad \beta_i = \frac{15 D_i b a_\infty}{R^2}. \tag{20}$$

By joining (18) to (15) and taking into account the smallness of a/a_∞ we obtain:

$$\frac{\partial a}{\partial t} \cong \frac{\beta_e \beta_i}{\beta_e + \beta_i}\left[C\left(1 - \frac{a}{a_\infty}\right) - \frac{1}{b}\frac{a}{a_\infty}\right]. \tag{21}$$

The influence of the longitudinal effects can approximately by accounted for in the mass-exchange coefficient if from system (19), (21) we go over to a system of equations of the hyperbolic type relating to some auxiliary dependent variables. However, since with sufficiently high x, $t\frac{a}{a_\infty} \ll 1$ and $\frac{\partial C}{\partial x} \ll \frac{\beta_э}{v} C$, we can go from (19), (21) over to system (1), (2) which is correct in this case only for the asymptotic stage of the process, moreover

$$\frac{1}{\beta_э} = \frac{1}{\beta_e} + \frac{1}{\beta_i} + \frac{D_э}{v^2}. \tag{22}$$

It can be shown that (16) remains the solution of this approximate system of equations. Thus the non-linear isotherm influences the forming stage for the stripe, while for the asymptotic stage, however far removed from the beingning of the process, the situation remains the same as developed at the transition stage.

For the boundary condition, when $p \to \infty$, that is for the linear isotherm of sorption, the quantity ba_∞ degenerates into the Henry coefficient, and instead of (16) the well-known equation for the Gauss curve can be obtained:

$$\varphi = 2\sqrt{\pi \frac{\Gamma}{\beta_э}} \frac{vs}{Q} C = e^{-z^2}. \tag{23}$$

Curve (23) is symmetrical and its peak corresponds to the condition $\xi = \tau$. The Henry constant Γ and the coefficient of mass-exchange can be derived from the following expressions:

$$\Gamma = \frac{t_m v}{x}, \quad \beta_э = 4\pi\left(\frac{v s t_m C_m}{Q}\right)^2 \frac{v}{x} \tag{24}$$

where:

C_m — is the concentration in the maximum of the stripe;
t_m — is the time of the appearance of the maximum behind the layer.

If $\beta_э$ is infinitely large, that is when the equilibrium chromatography occurs, it is easy to derive the following expressions from (14):

$$C = \frac{1}{b}\frac{\sqrt{ba_\infty \frac{x}{v}} - \sqrt{t}}{\sqrt{t}}, \quad ba_\infty \frac{x}{v} \geqq t \geqq ba_\infty \frac{x}{v}\left[1 - \sqrt{\frac{Q}{sxa_\infty}}\right]^2. \tag{25}$$

Thus, if the isotherm is convex the front is of an abrupt character. The place of the maximum of the stripe corresponds to the equality:

$$\sqrt{\frac{t_m}{ba_\infty}} = \sqrt{\frac{x}{v}} - \sqrt{\frac{Q}{vsa_m}} \tag{26}$$

while for the linear isotherm there is no second term in the right-hand part. Hence, the quantity ba_∞ that degenerates into the Henry coefficient if the concentrations are small, cannot be defined by the expression $ba_\infty = vt_m/x$, as it would lead to serious errors. The width of the stripe, as follows from (25) with regard to (26) is determined by the expression

$$C_m = \frac{1}{b} \frac{\sqrt{Q}}{\sqrt{sxa_\infty} - \sqrt{Q}}. \tag{27}$$

The point $C = 0$ in the tail-part of the stripe corresponds to the equality $\frac{1}{ba_\infty} t_{C=0} = \frac{x}{v}$, whence, with regard to (25), (26), we can easily obtain the following:

$$ba_\infty = \frac{vt_{C=0}}{x}, \qquad b = \frac{\sqrt{t_m} - \sqrt{t_{C=0}}}{C_m \sqrt{t_m}}. \tag{28}$$

Thus, in case of equilibrium chromatography, one curve is enough to obtain constants b and a_∞ and also to predict the development of the process of chromatography with any given length of the column. In case of non-equilibrium chromatography it is much more difficult to obtain the constants. The curves on Fig. 5 were calculated by (16) for various α. It is evident from Fig. 5 that

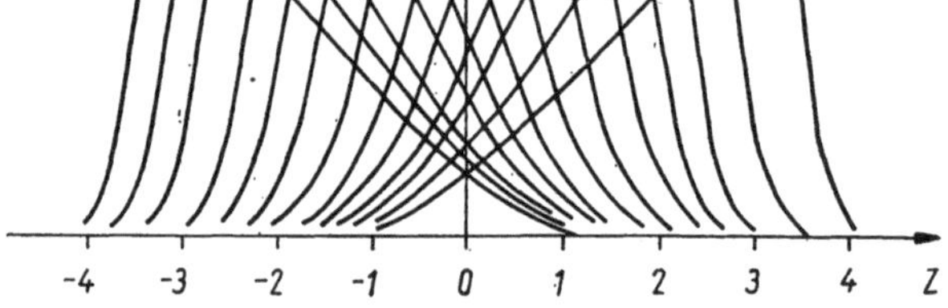

Fig. 5: Chromotographical stripes in the moving coordinate system.
α: 1 – −15; 2 – −12,5; 3 – −10; 4 – −7; 5 = −5; 6 – −3; 7 – −1; 8 – 7; 9 – 3; 10 – 5; 11 – 7; 12 – 10; 13 – 15.

as the absolute value of the complex α grows the height of the stripe maximum grows less, the maximum shifts more and more in respect to $z = 0$ and the asymmetry of the stripe increases. As these quantities are simply connected with one another it becomes possible to determine the unknown constants by some experimentally derived dimensionless criteria which define the degree of the asymmetry of the stripe. We can take as such a criterion the ratio of the area under the front part of the curve to the total area. Designating this ratio by η we obtain from (16)

$$\eta = \frac{1}{\alpha}\left[\ln(e^\alpha - 1) - \ln 2\sqrt{\pi}\, z_m - z_m^2\right] \tag{29}$$

and

$$\varphi_m = \frac{2\sqrt{\pi}}{\alpha} z_m . \tag{30}$$

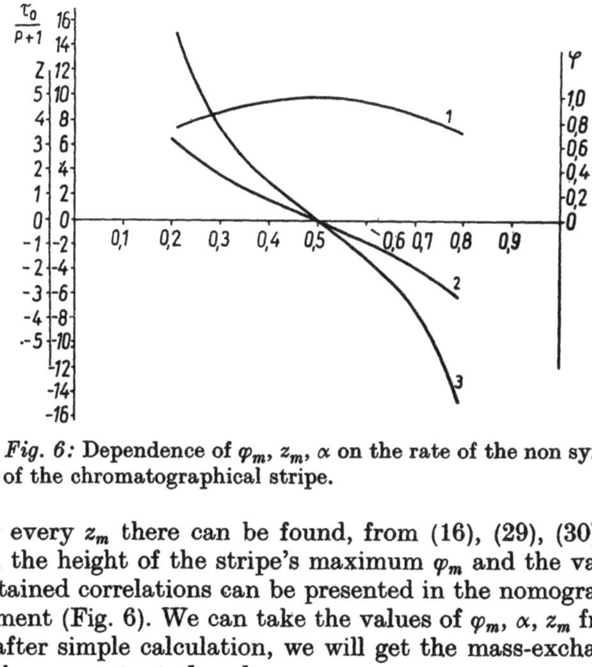

Fig. 6: Dependence of φ_m, z_m, α on the rate of the non symmetry η of the chromatographical stripe.

Thus, for every z_m there can be found, from (16), (29), (30) the degree of asymmetry, the height of the stripe's maximum φ_m and the value of α. After this, the obtained correlations can be presented in the nomogram form with η as the argument (Fig. 6). We can take the values of φ_m, α, z_m from this nomogram and, after simple calculation, we will get the mass-exchange coefficient and the isotherm constants b and a_∞.

$$b = \frac{z_m^2}{\alpha} \frac{Q}{C_m^2 v s t_m} \tag{31}$$

$$\beta_э = \frac{v}{x}\left(z_m + \frac{\alpha}{z_m} \frac{v s C_m t_m}{Q}\right)^2 \tag{32}$$

$$a_\infty = \frac{Q}{Sx\alpha}\left(z_m + \frac{\alpha}{z_m} \frac{v s C_m t_m}{Q}\right)^2 . \tag{33}$$

Having thus obtained the unknown constants for some chromatographic curve, we can make calculations beforehand for a curve, corresponding to some other quantity of substance introduced into the column, or to the different length of the column, or the different carrier-flow rate.

The above-examined laws of non-linear chromatography may make possible the quantitative substantiation of the optimum conditions for the separation of mixtures, with the use made of the discrepancies not only in the HENRY coefficient but in the curvature of the sorption isotherms also. The influence of this curvature on the chromatographic separation depends on the quantity of the substance introduced into the column, the temperature of the column and its length.

Suppose, for example, that the column sends forth almost simultaneously two stripes, one of which is symmetrical and the other, which is a little behind the first one, has a washed-out tail-end and a more steep front than the first one. Then we can draw the conclusion that the rate of movement of the asymmetrical stripe is less than that of the symmetrical one at the asymptotic stage and higher at the initial stage. As the formation-time of the stripe does not depend on the length of the column, then if the column is larger the distance between the maximums of the stripes will increase and the separation will be more complete. The same effect may be obtained if we lessen the flow-rate or the volume of the sample or if we increase the temperature.

The similar reasoning will be correct for any pair of components to be separated. The employment of the laws obtained for the practical chromatographical analysis is expedient not only when one substantiates the optimum conditions for the separation of substances, but also if the need arises to identify them.

3. The analysis of chromatographical processes based on the numerical solution of the dynamics problem

In the actual chromatographical processes there is usually a mixture of components undergoing the separation process and so a study of the dynamics of sorption for mixtures of substances is a rather important branch of research. However, when we go over from the sorption of one component to the theory of the processes of sorption for mixtures, the dynamics problem becomes considerably more complex, and even in the most simple instances, such as LANGMUIR's isotherm of sorption for the high rate of process, it becomes impossible to obtain an analytical solution for the dynamics problem. For the analysis of such processes the use of the numerical calculations obtained with IBM becomes essential. This circumstance considerably limits the possibilities for obtaining any general laws. But quantitative and qualitative laws can be examined in the most minute detail. The numerical solution of the problems of the dynamics of chromatographic processes may be obtained by the employment of various mathematical models — equilibrium and non-equilibrium, isothermal and non-isothermal ones. We shall limit ourselves to one of the simple models most frequently used for the analysis of chromatographic processes — the model of the equilibrium dynamics of sorption. In this model for each jth component the total combination of factors, to which the washing-out of the chromatographic stripe is due, is accounted for by a certain quasi-diffusional coefficient D_j^*.

We also assume that the equilibrium of sorption can be described by LANGMUIR's equation. For this case the mathematical model of the process is a system of non-linear equations in partial derivatives of hyperbolic type.

$$\begin{aligned} \frac{\partial C_j}{\partial t} &= D_j^* \frac{\partial^2 c}{\partial x^2} - v \frac{\partial C_j}{\partial x} - \frac{\partial a_j}{\partial t} \\ a_j &= \frac{A_j \Gamma_j C_j}{A_j + \sum_{k=1}^{n} \Gamma_k C_k}, \qquad n = 1, 2, \ldots \end{aligned} \tag{34}$$

Let the introduction of the sample in the chromatographic column be performed, as before, by a short impulse of duration. Then the initial and boundary conditions for the chromatographical process will be:

$$\begin{aligned} & t = 0, \qquad C_j = a_j = 0 \\ & t_0 > t > 0, \quad \begin{cases} x = 0, & vC_{0j} = vC_j(0) - D_j^* \dfrac{\partial C_j}{\partial x}; \\ x = H, & \dfrac{\partial C_j}{\partial x} = 0; \end{cases} \\ & t > t_0, \quad \begin{cases} x = 0, & vC_j(0) = D_j^* \dfrac{\partial C_j}{\partial x}; \\ x = H, & \dfrac{\partial C_j}{\partial x} = 0; \end{cases} \end{aligned} \tag{35}$$

where A_j and Γ_j — are the constants of the LANGMUIR equation and are determined from the sorption isotherms for the individual substances; D_j^* — the coefficient, which accounts for the total combination of effects, due to which the washing-out of the chromatographic stripe for the "j" component occurs. For the convenience of the following analysis of the results we introduce the dimensionless parameters and variables in accordance with the correlations:

$$\xi = \frac{x}{H}; \quad y_j = \frac{C_j}{C_{0j}}; \quad \gamma_j = \frac{a_j}{a_{0j}}; \quad \tau = \frac{vtC_{01}}{HA_1}$$

$$G_j = \frac{D_j^*}{Hv}; \quad j = 1, 2, \ldots, n.$$

So, in a dimensionless form, the system of differential equations (34) and boundary and initial conditions (35) appear in the following form:

$$\frac{C_{01}}{A_{01}}\frac{\partial y_j}{\partial \tau} + \frac{A_j}{A_1}\frac{C_{01}}{C_{0j}}\frac{\partial \gamma_j}{\partial \tau} + \frac{\partial y_j}{\partial \xi} = G\frac{\partial^2 y_j}{\partial \xi^2}$$

$$\gamma_j = \frac{y_j \varphi_j}{P_j}; \quad \varphi_j = \frac{1}{1 + \sum_{k=1}^{n} \frac{y_k A_k}{P_k A_j}}; \quad P_j = \frac{A_j}{\Gamma_j C_{0j}}$$

$$\tau = 0, \qquad y_j = \gamma_j = 0 \tag{36}$$

$$\tau_0 > \tau > 0, \begin{cases} \xi = 0, & 1 = y_j(0) - G_j \dfrac{\partial y_j}{\partial \xi}; \\ \xi = 1, & \dfrac{\partial y_j}{\partial \xi} = 0; \end{cases}$$

$$\tau > \tau_0, \begin{cases} \xi = 0, & y_j(0) = G_j \dfrac{\partial y_j}{\partial \xi}; \\ \xi = 1, & \dfrac{\partial y_j}{\partial \xi} = 0. \end{cases}$$

For a numerical solution of such systems of equations the indefinite differential schemes are the most effective. In this case the authors used the following end-differences approximation of the initial system of equations:

$$\mu_1(y_{ji}^{k+1} - y_{ji}^k) + \frac{\mu_1}{\mu_j P_j}\frac{y_{ji}^{k+1}\varphi_{ji}^{k+1} - y_{ji}^k \varphi_{ji}^k}{l} +$$

$$+ \frac{y_{jk+1}^{k+1} - y_{ji}^{k+1}}{h} = G_j \frac{y_{ji+1}^{k+1} - 2y_{ji}^{k+1} + y_{ji-1}^{k+1}}{h^2};$$

$$\frac{C_{0j}}{A_j} = \mu_j; \quad A_j = \frac{1}{h} - \frac{G_j}{h^2}; \quad C_j = \frac{-G_j}{h^2};$$

$$B_j = \frac{1}{2}\left(\frac{1}{h} - \frac{2G_j}{h^2} - \frac{\mu_1}{l} - \frac{\mu_1 \Gamma_0 \varphi_{ji}^{k+1}}{l}\right);$$

$$D_j = \frac{\mu_1}{l} + \frac{\mu_1 \Gamma_j}{l}\varphi_{ji}^k;$$

$$A_j y_{i+1}^{k+1} - 2B_j y_i^{k+1} + C_j y_{i-1}^{k+1} = D_j y_i^k.$$

When working out the algorithm we understood the solution of the problem of the dynamics of the chromatographical process as the determination of the distribution of concentrations $C_j(x, t)$, $a_i(x, t)$. In order to evaluate the changes in the forms of the distribution curves for various components of the mixture during their movement along the column, and in order to determine the chromatographic stripes, for the solution of the problem we calculated the correlations $f_j = \frac{S_{ij}}{S_{j\text{total}}}$ where S_{ij} — the area of the stripe to the left of the maximum concentration (the quantity of substance in the rear part of the stripe) $S_{j\text{total}}$ — the total area of the chromatographic stripe (the quantity of substance in a stripe limited by concentration $C_j \geqq \ldots$)

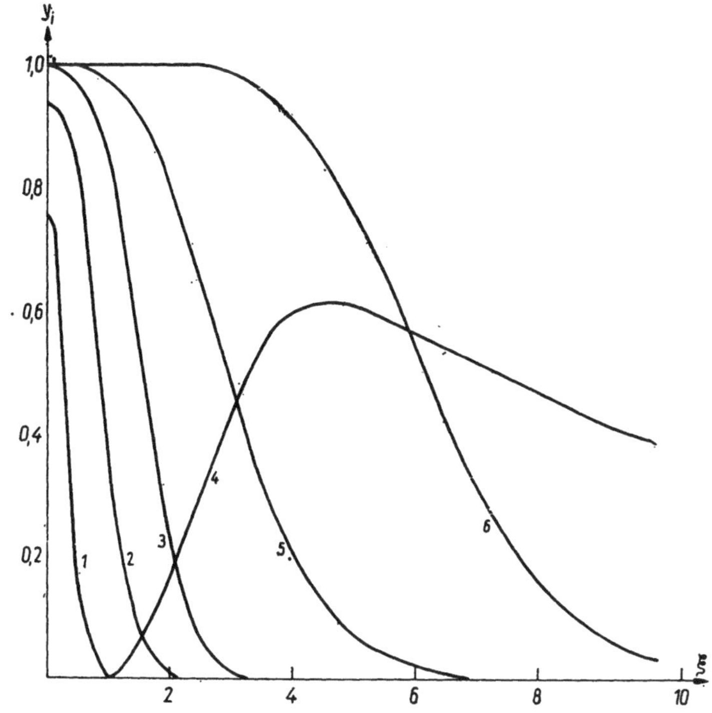

Fig. 7: Initial stages of process of the chromatographical separation of the two-component mixture.

$C_1 = 1 \cdot 10^{-6}$ g/cm³; $C_2 = 0.1 \cdot 10^{-6}$ g/cm³;
$v = 2.81$ cm/s; $g_j = 0.02$; $H = 5$.

Here we give some examples of the results of the separation process for binary and three-component mixtures. Fig. 7 gives the beginning stage of the formation of the chromatographic stripe for the separation of a binary mixture.

As can be observed in Fig. 8, the chromatographic stripes become considerably deformed with the movement of the components through the layer. The stationary conditions are formed during a lengthy period of time that depends on the parameter P_i and on the volume of the sample introduced into the column.

Fig. 9 gives the chromatographic stripes for various lengths of the sorbent layer in the column. As the chromatographic stripes wash out with the movement along the column the maximum concentrations of the components in the stripe decrease and asymptotically approach some quantites which can be determined by equilibrium and kinetic parameters (Fig. 10).

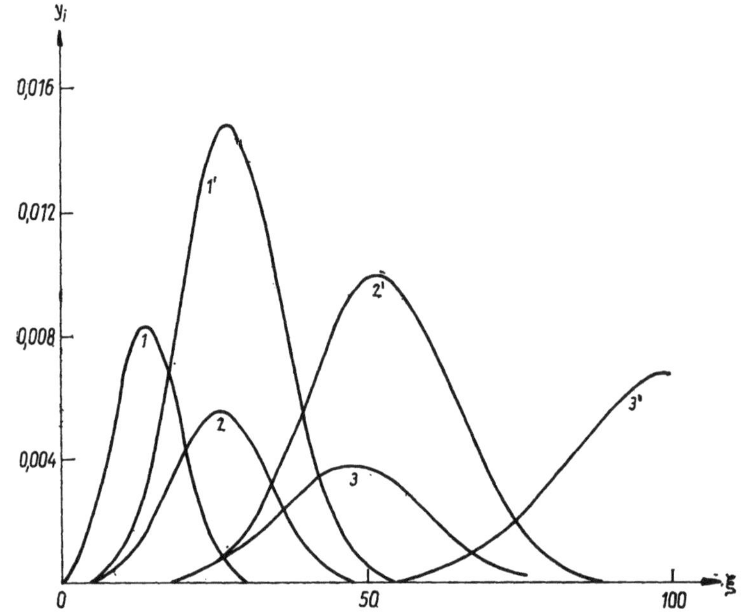

Fig. 8: Process of the chromatographical separation of the two component mixture.

$C = 1 \quad 10^{-6}$ g/cm³; $C_2 = 0.1 \cdot 10^{-6}$ g/cm³;
$v = 2.81$ cm/s; $g_j = 0.02$;
1, 2, 3 – 1st component;
1, 2, 3 – 2nd component;
1.1' – $\tau = 1968$ min; 2,2' – $\tau = 3937$ min;
3.3' – $\tau = 7884$ min.

Table 1 and Fig. 11, 12 show the results of the separation process for a three-component mixture obtained by means of an IBM model.

With the changes in D_1^* (exp. 3, 4) only $C_{1\max}$ and $t_{1\max}$ grow less, while they stay the same for the second and the third components. Similarily, when D_2^* grows from 2.8 to 20 (exp. 4—6) then thore is decrease only in the characteristics of. the chromatographic stripes for the second component. It is shown by the results in Fig. 11 that the degree of wash-out of the chromatographic stripes for the less absorbable (2nd, 3rd) components grows gradually less from

Table 1: The calculations for the separation of a three-component Mixture

NN	H	D_j^*, cm^2/s			$C_{j\max}$			Time, $t_{j\max}$, min		
	cm	1	2	3	1	2	3	1	2	3
1	30	2.8	2.8	2.8	0.077	0.908	0.967	107	7.6	6.4
2	60	2.8	2.8	2.8	0.056	0.703	0.802	221	12.5	9.8
3	100	10.0	2.8	2.8	0.022	0.523	0.608	347	19.1	14.8
4	100	20.0	2.8	2.8	0.0168	0.522	0.608	316	19.1	14.8
5	100	2.8	10.0	2.8	0.0486	0.385	0.608	375	18.0	14.8
6	100	2.8	20.0	2.8	0.0486	0.319	0.608	375	16.9	14.8
7	100	2.8	2.8	10.0	0.0486	0.522	0.479	375	19.1	14.2

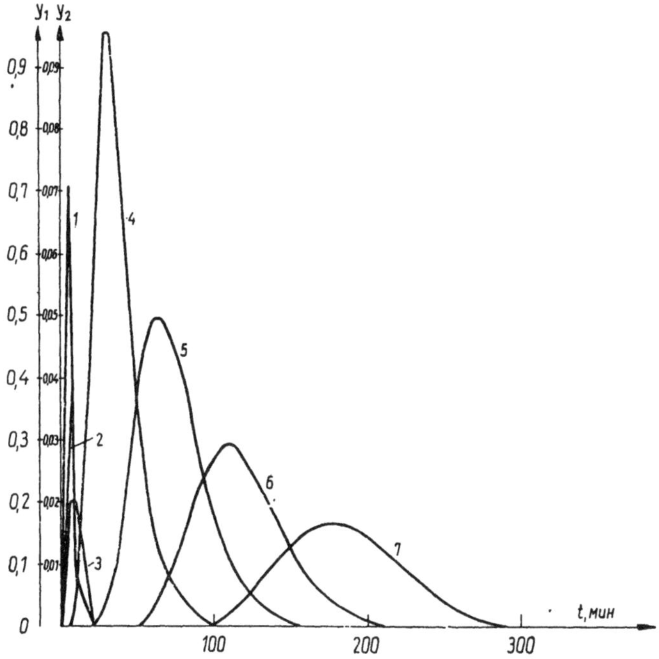

Fig. 9: Dependence of y_j on the length of the column.

$C_1 = 1.0\ 10^{-6}$ g/cm³; $C_2 = 0.1 \cdot 10^{-6}$ g/cm³;
$v = 2.81$ cm/s; $g_j = 0.2$;
1, 2, 3 — 1st component;
4, 5, 6, 7 — 2nd component;
4 — $H = 10$ cm;
1.5 — $H = 20$ cm;
2.6 — $H = 32$ cm;
3.7 — $H = 50$ cm.

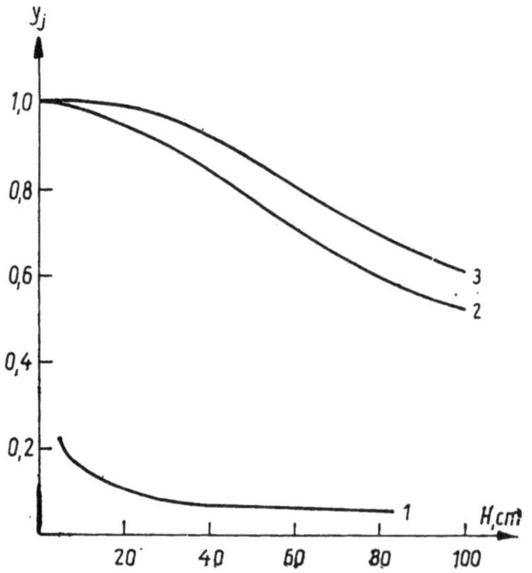

Fig. 10: Dependence of y_j on the length of the column.
$C_1 = 1.0 \cdot 10^{-6}$ g/cm³; $C_2 = 0.1 \cdot 10^{-6}$ g/cm³;
$C_3 = 0.05 \cdot 10^{-6}$ g/cm³;
1 — 1st component Y_1;
2 — 2nd component Y_2;
3 — 3rd component Y_3.

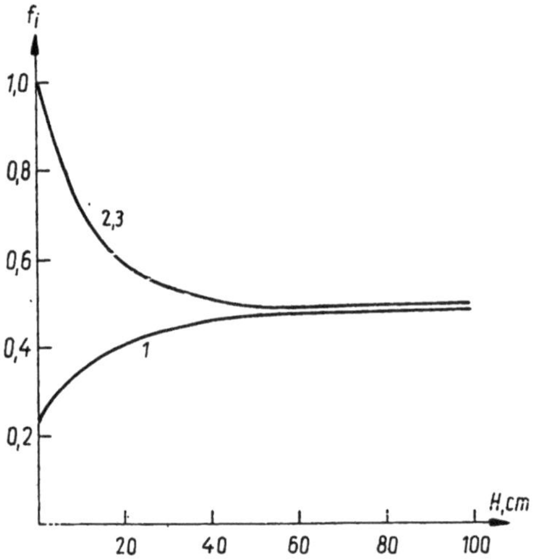

Fig. 11: Dependence of f_j on the length of the column.
$C_1 = 1 \cdot 10^{-6}$ g/cm³; $C_2 = 0.1 \cdot 10^{-6}$ g/cm³;
$C_3 = 0.05 \cdot 10^{-6}$ g/cm³;
1 — 1st component;
2 — 2nd component;
3 — 3rd component.

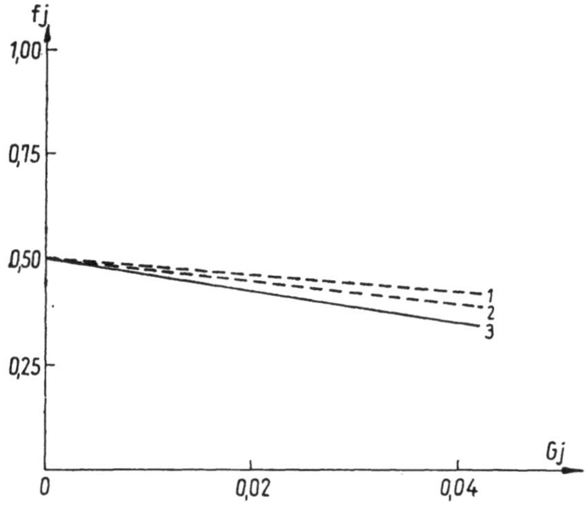

Fig. 12: Dependence of f_j on D_j.

$C_1 = 1 \cdot 10^{-6}$ g/cm³; $C_2 = 0.1 \cdot 10^{-6}$ g/cm³;
$C_3 = 0.05 \cdot 10^{-6}$ g/cm³;
1 — 1st component;
2 — 2nd component;
3 — 3rd component.

the maximum down to $f_j \approx 0.5$, which corresponds to the symmetrical stripe. At the same time the coefficient of asymmetry for the most absorbable 1st component rises asymptotically up to a value of 0.5 if the layer is sufficiently long. The longitudinal wash-out effects (Fig. 12) lessen the values of f_j in accordance with the absorbability of the components in the mixture.

The numerical calculations with IBM allow one to determine the separability of the components of a mixture and their absorbable volumes, to calculate the values of maximum concentrations of components in a chromatographic stripe, to choose the conditions for the experiment that allow for the set quality of separation. Physicochemical constants, which characterise the absorbability of the components of a mixture and wash-out of the chromatographic stripes (for the above described mathematical model $A_{\infty j}$, Γ_j, D_j^*) can be determined from the experimental data, by employment of the known methods with unlinear programs.

4. The chief designations

x, t — current co-ordinate and time
C — current concentrations of the sorbate in a moving phase
a_∞. — maximum value of sorption

Γ, b — constants in the equation for the isotherm of sorption
v — linear flow-rate
ε — porosity of the layer
$\beta_э$ — coefficient of mass-transfer
r, R — distance from the heart of the grain and its radius
D_i — coefficient of inner diffusion
$D_э$ — coefficient of longitudinal transfer
a_i, $a_з$, a — local, surface and average integral over the grain volume values of sorption.

Mathematical modelling of the dynamics of the adsorption processes with reverse connection of concentrations

(V. L. Kolin, E. A. Ustinov, Y. N. Taganov, A. V. Kolbancev, A. K. Akulov, A. A. Seballo)

Zusammenfassung:

Es wird ein mathematisches Modell für die adsorptive Reinigung eines Arbeitsmediums von Verunreinigungen, das in einem geschlossenen Raum zirkuliert, vorgeschlagen. Lösungsmethoden für diese Aufgabe werden untersucht. Es werden numerische, aber auch für einige spezielle Fälle analytische Lösungen angegeben. Es wird gezeigt, daß wegen der Instationarität des Adsorptionsprozesses im Festbett und der Reversibilität der Adsorption in solchen Systemen oszillierende Rückkopplungen bezüglich der Sorptivkonzentration auftreten, die zu periodischen Konzentrationsschwingungen im geschlossenen Raum führen können. Bei der Adsorption von Gemischen kann die Verdrängung einer Komponente durch die andere zu einer periodischen Konzentrationserhöhung dieser Komponente im Vergleich zu seiner Anfangskonzentration im geschlossenen Raum führen.

Es wird der Einfluß verschiedener Faktoren auf die Amplitude und Frequenz der Konzentrationsschwingungen untersucht. Die Rechenergebnisse werden mit experimentellen Daten verglichen.

1. Introduction

In various fields of technology the working media (gaseous or liquid) circulating in a closed circuit are cleared of the impurities, either existent in the media or coming in from outside, by means of sorption processes. The typical examples of such a use of adsorption are air-conditioning for hermetically closed living-quarters either on earth, in space or under water, air-conditioning in ware-houses where fruits or vegetables are stored, purification of working media in various apparatuses where the media are polluted by harmful products. Sorption is also widely used in closed technological water circuits and in surgery (reconstructive systems, e.g. artificial kidney).

For all these cases the concentration of the impurity in the working media changes according to a low, which depends on the rate at which the impurity enters the system, on the velocity of the circulation of the media through the sorption filter, on the interfusion of the floods in the vessel and on the adsorption factor.

In spite of the great importance of these processes and their wide practical employment, the theory of the process and methods of calculation are not yet fully developed.

Every singl process has its peculiar features, which are the quantity, composition and sorbability of the impurities, the rate at which they are being produced, the demands on the quality of the purification, characteristics of stirring up in a hermetic vessel, measurements of the adsorption filters, the velocity of the flow through the filter, etc.

However, undoubtedly the main characteristic feature of these processes, which distinguishes them from the more thoroughly investigated processes under frontal, eluent or developing conditions, is the two-way connection between the concentration of the sorbate on its entering the filter and on its eaving the filter.

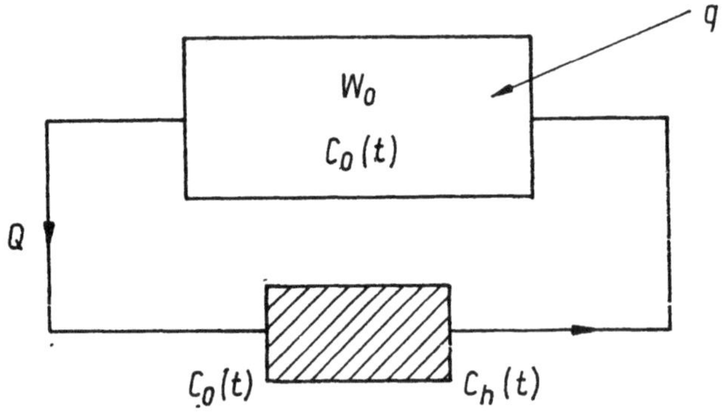

Fig. 1: General scheme.

Therefore, for the study of the general law of the process of sorption with the two-way connection of the concentrations we shall confine ourselves to the study of the scheme in Fig. 1. For this study we assume that in volume W_0, where the two-way connection takes place, there is complete stirring up of the working media.

For most technological processes this assumption is close to the true structure of the flow in the working volume.

2. The dynamics of processes with linear isotherm

In order to investigate the basic laws of the process we will first look into some characteristic individual cases: equilibrium and non-equilibrium dynamics of isothermal sorption with linear isotherm.

Bearing in mind the assumption for the structure of the flow in volume W_0 we will write the equation of material balance as follows

$$\frac{\partial C_0}{\partial t} = \frac{a}{W_0} - \frac{Q}{W_0}(C_0 - C_h). \tag{1}$$

If the process of sorption proceeds under non-isothermal conditions and the influence of the effects of longitudinal transfer is not strong, then the mathematical model of the process of sorption conforms to the following system of equations in partial derivatives of hyperbolic type:

$$\varepsilon \frac{\partial c}{\partial t} = -v \frac{\partial c}{\partial x} - (1 - \varepsilon) \frac{\partial a}{\partial t} \tag{2}$$

$$\frac{\partial a}{\partial t} = \beta_{\text{eff}} \left(c - \frac{a}{G} \right) \tag{3}$$

with the initial and boundary conditions:

$$a(x, 0) = 0, \quad c(0, t) = C_0, \quad C_0(0) = C_{00}. \tag{4}$$

Where:

x, t — current coordinate, taken from the moment of entrance of the flow into the filter, and time;
$c; a$ — current concentrations of the sorbate within the filter;
v — linear velocity of the flow through the complete cross-section of the filter;
c_0 — concentration of the sorbate in volume W_0;
Q — the flow rate of the media through W_0;
C_h — concentration of the sorbate as it leaves the filter;
q — the rate of incoming of the impurity into the volume;
G — the HENRY constant;
ε — porosity of the sorbent layer;
β_{eff} — accounts for purely kinetic factors and for the effects of longitudinal transfer, $\beta_{\text{eff}} = k_D \cdot \beta_0$, where β_0 — kinetic coefficient,

$$k_D \in [k_{D_1}, k_{D_2}]; \quad k_{D_1} = \frac{2}{1 + \sqrt{1 + 4D}}; \quad k_{D_2} = \frac{1}{1 + D}; \quad D = \frac{\beta_0 \cdot D_e}{v^2}$$

D_e — effective coefficient for longitudinal transfer.

In cases of equilibrium sorption ($\beta_{\text{eff}} \to \infty$) instead of (3) we will have the isotherm equation $a = Gc$. If $\beta_{\text{eff}} \to \infty$, then the "instantaneous" slip-through concentration $C_h = 0$.

This conforms to the tear-off front of concentrations at $x = h$. As in our case $\frac{da}{dc} = G$ then, according to the law by WICKE [1], the velocities of all concentration points of the front of sorption are equal at any moment and the rate of movement of the tear-off front is determined by the equation by WILSON [2]:

$$u = \frac{v}{1 + G}. \tag{5}$$

Concentration of the sorbate at $x = 0$ is variable, and $u = \text{const}$ so the function $C_h\left(\frac{h}{v}\right)$ will follow the changes of $C_0(t)$ though being a little late. The value of the lateness Δt depends on the time which is needed by the front of sorption in order to cover the distance $x = h$, that is $\Delta t = (h/v) \cdot (1 + G)$.

The slip-through concentration of the sorbate in leaving the layer, which is h long, is nil at the beginning of the process. Hence by integrating (1) from 0 to Δt we can obtain $C_0(t)$ up to the moment when the tear-off front leaves the filter.

For the next stage $\Delta t \leqq t \leqq 2\Delta t$, $C_h > 0$, but it will change in the same way as the one obtained by integrating (2) $C = C_0(t)$, where $0 \leqq t \leqq \Delta t$.

This allows to repeat the integration of (2) for $\Delta t \leqq t \leqq 2\Delta t$ and so on.

Thus, for this case, the problem of the dynamics of sorption (for purification) is reduced to the solution of the equation

$$\frac{dy_0}{d\tau} + k(d_0 - d_H) - k_1 = 0, \tag{6}$$

where

$$y_0 = \frac{C_0}{C_{00}} \quad y_H = \frac{C_h}{C_{00}}, \quad \tau = \frac{v \cdot t'}{h(1+G)}, \quad t' = t - \frac{x}{v}\varepsilon \simeq t,$$

$$k = \frac{(1+G)Wf}{W_0}, \quad k_1 = \frac{(1+G)Wf \cdot q}{Q \cdot W_0 \cdot C_{00}}, \tag{7}$$

$\tau = 1$ — the time, in which the front covers the distance $x = h$. As, if $0 \leqq \tau \leqq 1$, then $y_H = 0$, from (6) we obtain

$$\frac{dy_0}{d\tau} = -k\left(y_0 - \frac{k_1}{k}\right). \tag{8}$$

And, integrating (8)

$$y_0 = \frac{k_1}{k} + B_0 \exp(-k\tau), \tag{9}$$

B_0 — const. $y_0(0) = 1$,

hence $B_0 = 1 - \frac{k_1}{k}$, and

$$y_0 = \frac{k_1}{k} + \left(1 - \frac{k_1}{k}\right)\exp(-k\tau), \qquad 0 \leqq \tau \leqq 1 \tag{10}$$

Equation (10) defines y_H for the time interval $1 \leqq \tau \leqq 2$

By introducing y_H thus obtained into (6) we will have the equation which describes how y_0 varies when $1 \leqq \tau \leqq 2$.

$$\frac{dy_0}{d\tau} + ky_0 - k\left[\frac{k_1}{k} + \left(1 - \frac{k_1}{k}\right)\exp\left(-k(\tau - 1)\right)\right] - k_1 = 0. \tag{11}$$

When we replace $\frac{dy_0}{d\tau}$ by $\frac{dy_0}{d(\tau - 1)}$ and, instead of (11)

$$\frac{dy_0}{d(\tau - 1)} + ky_0 - k\left(1 - \frac{k_1}{k}\right)\exp\left[-k(\tau - 1)\right] - 2k_1 = 0. \tag{12}$$

After integrating (12) we have

$$y_0 = 2\frac{k_1}{k} + \left(1 - \frac{k_1}{k}\right) k(\tau - 1) \exp\left[-k(\tau - 1)\right] + B_1 \exp\left[-k(\tau - 1)\right]. \quad (13)$$

When $\tau = 1$ y_0 does not have a breach, so when $\tau = 1$ the values of y_0 obtained from (9) and (13) should be equal and thus $B_1 = \left(1 - \frac{k_1}{k}\right) \exp(-k) - \frac{k_1}{k}$.

Hence

$$y_0 = 2\frac{k_1}{k} + \left(1 - \frac{k_1}{k}\right) e^{-k\tau}[e^k(\tau - 1)\, k + 1] - \frac{k_1}{k} e^{-k\tau} e^k, \quad (14)$$

$$1 \leqq \tau \leqq 2.$$

By similar reasoning we derive the general law

$$y_0 = (n + 1)\frac{k_1}{k} + e^{-k\tau}\left\{\left(1 - \frac{k_1}{k}\right) \sum_{s=0}^{n} \frac{k(\tau^s - S)^s}{S!} - \frac{k_1}{k} \sum_{m=0}^{u} e^{mk} \sum_{s=0}^{m-l} \frac{k^s(\tau - m)^s}{S!}\right\}, \quad n \leqq \tau \leqq n + 1. \quad (15)$$

From (15) various individual cases may easily be derived. Thus, when there is no source of impurities ($k_1 = 0$), then, from (15)

$$y_0 = e^{-k\tau} \sum_{s=0}^{n} \frac{k^s(\tau - s)^s}{S!} e^{sk}, \quad n \leqq \tau \leqq n + 1 \quad (16)$$

Calculations made by (15) and (16) show that the concentration in the system changes non-monotonously and that the case in hand is characterised by fading oscillations (Fig. 2).

If $k_1 = 0$ the value of y_0 keeps swaying round 0.2. When $k = 4$ as in our calculations, then $y_0 = 0.2$ conforms to the adsorption equilibrium between the concentration of adsorbate in the moving media (gaseous, liquid) and the adsorbent.

In case when the mass velocity of the impurity coming into volume W_0 remains the same, then the concentration y_0 first monotonously lessens, as when $k_1 = 0$, then begins to increase. In both these cases the amplitude of the oscillations lessen in time while the period stays constant and is determined by the time, which it takes the front to cover the distance $x = h$. Hence, the greater the length of the filter h, the greater G and the less v, the longer will be the period and the less the amplitude of the oscillations, that is the oscillatory process becomes less defined.

When $k_1 = 0$ the amplitude of the first oscillation of the concentration h_1 (the count begins from the first minimum of concentrations) is derived from

$$h_1 = e^{-2kH} + (kH - 1)\, e^{-kH}.$$

As is shown by analysis, the dependence of the amplitude of oscillations of the concentration of impurity in the volume W_0 on kH is non-monotonous. The oscillations vanish when $kH \to 0$ as well as when $kH \to \infty$, that is when the adsorption capacity of the bed is either very small or very large.

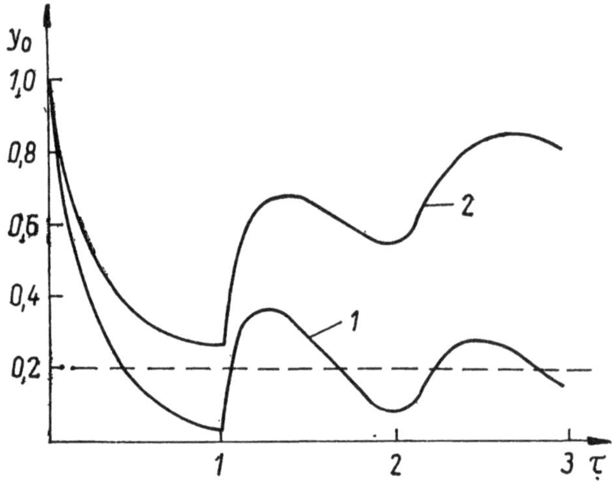

Fig. 2: Cleaning dynamics in the comprised volume. Equilibrium sorption.
$k = 4$; k1: 1 – 0; 2 – 1.

The maximum amplitude of oscillations corresponds to $kH \simeq 1.5$.

Now let us consider the non-equilibrium process. Ignoring the difference between t and $t' = t - \frac{h}{v}\varepsilon$ we can rewrite the system (1)–(14) in the dimensionless form as follows

$$\frac{dy_0}{d\tau} = k_1 - k(y_0 - y_H) \tag{17}$$

$$\frac{\partial^2 y}{\partial\tau\,\partial\xi} + \frac{\partial y}{\partial\xi} + \frac{\partial y}{\partial\tau} = 0 \tag{18}$$

$$\tau = 0, \quad \xi \geqq 0, \quad y = \exp(-\xi); \quad \tau \geqq 0, \quad \xi = 0 \quad y = y_0, \quad y_0(0) = 1 \tag{19}$$

where

$$\xi = \frac{\beta_{\text{eff}} \cdot x}{v}(1 - \varepsilon); \quad \tau = \frac{\beta_{\text{eff}}}{G} t'; \quad y_0 = \frac{C_0}{C_{00}};$$

$$y = \frac{C}{C_{00}}, \quad y_H = \frac{C_h}{C_{00}}; \quad k = \frac{Q \cdot G}{W_0 \cdot \beta_{\text{eff}}}; \quad k_1 = \frac{q \cdot G}{W_0 \beta_{\text{eff}} \cdot C_{00}}.$$

For the solution of this problem we will take the LAPLACE transformation over the variable τ. As regards the representations the initial system takes the form:

$$-1 + p\bar{y}_0 + k\bar{y}_0 - k\bar{y}_H - k_1 \cdot \frac{1}{p} = 0 \tag{20}$$

$$\bar{y} = A \cdot \exp\left(-\frac{p}{1+p} \cdot \xi\right) \tag{21}$$

where A — the integration constant and $A = \bar{y}_0$, hence

$$\bar{y}_H = \bar{y}_0 \exp\left(-\frac{p}{1+p} H\right), \quad H = \frac{\beta_{\text{eff}} \cdot h}{v} (1 - \varepsilon). \tag{22}$$

Excluding $\bar{y}_H$ from (20), (22) we will find

$$\bar{y}_0 = \left(\frac{1}{p} + \frac{k_1}{p^2}\right) \Big/ \left[1 + \frac{k}{p}\left(1 - \exp\left(-\frac{pH}{1+p}\right)\right)\right] \tag{23}$$

and back to the original

$$y_0 = e^{-k\tau} \sum_{n=0}^{\infty} \frac{\tau^n}{n!} \sum_{m=0}^{n} \frac{H^m(1-k)^{n-m}}{m!} \cdot \sum_{s=0}^{n-m} \frac{Z^s \cdot S^m(n-m)!}{S!(n-m-S)!}$$
$$+ k, \; e^{-\tau} \sum_{n=1}^{\infty} \frac{\tau^n}{n!} \sum_{i=0}^{n-1} \frac{H^m(1-k)^{i-m}}{m!} \sum_{s=0}^{i-m} \frac{Z^s \cdot S^m(i-m)!}{S!(n-S-i)!}, \tag{24}$$

where $\quad Z = k_3/k_2;\ k_2 = 1 - k;\ k_3 = \exp(-H)$.

The formula (24) describes the law, by which the concentration varies in volume W_0 for all stages of the purification process.

The expression (24) is too cumbersome and to interpret it is not easy even with a digital computer. In order to analyse general regularities of the process we will consider the borderline cases: purification of the working medium from impurities at the initial stages of the process ($\tau \to 0$) and when the values of τ are sufficiently high.

Research shows that for initial and asymptotic stages of the process simplier approximate correlations than (24) could be obtained. Thus for $\tau \to 0$, $p \to \infty$

$$\bar{y}_0 \simeq \frac{1}{1+p} + \frac{k_1}{p(p+k)}. \tag{25}$$

Hence follows the expression describing the change in the concentration in the volume W_0 at the initial stage of the process of adsorption

$$y_0 \simeq \left(1 - \frac{k_1}{k}\right) e^{-k\tau} + \frac{k_1}{k}, \quad \tau \to 0. \tag{26}$$

The expression (26) is correct while the adsorbate "slip-through" beyond the sorbent layer h long is rather small and does not influence the change in the concentrations in the volume W_0. Solution (26) may also be obtained by direct integration of the material balance equation (15) on the condition $y_H = 0$. The latter takes place either at the initial stage of the process of adsorption or if the volume of the filter is infinitely large. At the asymptotic stage, .e. when $\tau \to \infty$, $p \to 0$ the expression $\frac{1}{p}\left[1 - \exp\left(-\frac{pH}{1+p}\right)\right]$ will become ndeterminate. Opening the indeterminacy by L'HOSPITAL's rule we get:

$$\frac{1}{p}\left[1 - \exp\left(-\frac{pH}{1+p}\right)\right] \simeq \frac{H}{(1+p)^2} \cdot \exp\left(-\frac{pH}{1+p}\right). \quad (27)$$

Therefore the expression (23) assumes for high values of τ the following form

$$\bar{y}_0 = \left(\frac{1}{p} + \frac{k_1}{p^2}\right) \Big/ \left[1 + \frac{k \cdot H}{(1+p)^2} \cdot \exp\left(-\frac{pH}{1+p}\right)\right], \quad (28)$$

or

$$y_0 \simeq \left(\frac{1}{p} + \frac{k_1}{p^2}\right)\left[1 - \frac{kH}{(1+p)^2} \cdot e^{-\frac{pH}{1+p}} + \frac{(kH)^2}{(1+p)^4} e^{\frac{-2pH}{1+p}} - \cdots\right]. \quad (29)$$

The value of $p < 1$, hence it is possible to simplify the denominators of the pre-exponental factor in the expression (29):

$$\bar{y}_0 \simeq \left(\frac{1}{p} + \frac{k_1}{p^2}\right)\left[1 - kHe^{-\frac{pH}{1+p}} + (KH)^2 \cdot e^{-\frac{2pH}{(1+p)'}} - \cdots\right]. \quad (30)$$

Hence, after transition to the original we will obtain the expression for the determination of the sorbate concentration in the volume W_0 when the values of τ are sufficiently high:

$$\begin{aligned} y_0 \simeq{} & 1 - kH \cdot y_l(H, \tau) + (kH)^2 y_l(2H, \tau) - \cdots \\ & + k_1[\tau - kHy_{lk}(H, \tau) - (kK)^2 y_{lk}(2H, \tau) - \cdots]. \end{aligned} \quad (31)$$

In the expression (31) $y_l(mH, \tau)$ — is the function of distribution of the concentration y over the sorbate layer in time for the frontal sorption process; $y_{lk}(mH, \tau)$ is the cumulative function for the same process.

These functions and the asymptotic approximations to them have, correspondingly, the following form

$$y_l(\xi, \tau) = e^{-(\xi+\tau)} \sum_{n=0}^{\infty} \frac{\xi_n}{n!} \sum_{m=n}^{\infty} \frac{\tau^m}{m!} \quad (32)$$

$$y_{lk}(\xi, \tau) = e^{(\xi+\tau)} \sum_{n=0}^{\infty} \frac{\xi^n}{n!} \sum_{m=n+1}^{\infty} \frac{(m-n) \cdot \tau^m}{m!} \quad (33)$$

$$y_l(\xi, \tau) \simeq \frac{1}{2} \operatorname{erfc}\left(\sqrt{\xi} - \sqrt{\tau}\right) + \frac{e^{-(\sqrt{\xi}-\sqrt{\tau})^2}}{4\sqrt{\pi}\,(\xi \cdot \tau)^{1/4}} \quad (34)$$

$$y_{lk}(\xi. \tau) \simeq \frac{1}{2}(\tau - \xi) \operatorname{erfc}\left(\sqrt{\xi} - \sqrt{\tau}\right) - \frac{(\sqrt{\xi} - \sqrt{\tau})^2 \cdot e^{-(\sqrt{\xi}-\sqrt{\tau})^2}}{4\sqrt{\pi}\,(\xi \cdot \tau)^{1/4}}. \quad (35)$$

Thus the use of superposition of the solution for the problem of the dynamics of frontal sorption with a linear isotherm allows for the calculation of the process of purification of the working medium in a close-circuit system.

It should be noted that members of series (31) increase from zero to a certain limit not simultaneously but one after another, while the increase for each preceding member goes more slowly than for the subsequent one.

Thus for the region of values of τ comparable to H, the function y_0 decreases, in the region $\tau \simeq 2H$ increases, then decreases again and so on. With the increase in τ the separate component functions superpose and the oscillations of the function y_0 smooth down.

When $k_1 = 0$, i.e. no impurities are introduced into the system there should arise an equilibrium condition for the borderline case of $\tau \to \infty$. It is obvious that the amplitude of oscillations is the higher the higher H and kH are, i.e. the less are Q and W_0 as compared to the volume of the adsorbent filter. The non-equilibrium state of adsorption leads to the washing-out of the concentration waves not only inside the adsorbent layer but also in the working volume W_0.

The examples given and analysed above show that the analytical solutions for the problems of dynamics of adsorption processes with reverse communication by concentration are complex; to obtain them and put them to any practical use is difficult even in the case of most simple linear models.

All the same, the linear isotherm of sorption is an important borderline case, which allows in a first approximation to describe adsorption for the region of low concentrations, and linear models offer the opportunity to study and further calculate the processes of profound purification of the working medium in a closed volume.

When solving systems of linear differential equations with partial derivatives by the operator method, the basic difficulty lies as a rule in finding the original of a function from its image.

For this purpose the numerical method could be employed. We will approximate the original of the function y by a polynomial with the powers $\alpha = \exp(-h\tau)$, where h is a constant. By the definition of the LAPLACE transformation it is obvious that

$$\bar{y}(p) = \int_0^\infty e^{-p\tau} y(\tau) \cdot d\tau \tag{36}$$

where $\bar{y}(p)$ is the image of the original $y(\tau)$, p — transformation parameter. If $p = (k + 1)\, h$ instead of (36) we can write

$$h\bar{y}[(k + 1)\, h] = \int_0^1 \alpha^k y(\alpha)\, d\alpha \tag{37}$$

or

$$b_n = \sum_{k=0}^{n} \frac{(-1)^{n+k}\,(n + k)!}{(n - k)!\,(k!)^2}\, h \cdot \bar{y}[(k + 1) \cdot h] = \int_0^1 P_n(\alpha)\, y(\alpha)\, d\alpha \tag{38}$$

where $P_n(\alpha)$ is LEGENDRE's polynomial of n order.

As

$$\int_0^1 P_n \cdot P_k \cdot d\alpha = \begin{cases} 0, & k \neq n \\ \dfrac{1}{2n+1}. & k = n \end{cases}$$

then confining ourselves for engineering purposes to m power of the polynomial approximating the original we get

$$y(\alpha) \simeq \sum_{n=0}^{m} (2n+1)\, b_n \cdot F_n(\alpha) \tag{39}$$

or

$$y(\alpha) \simeq \sum_{n=0}^{m} a_n \cdot h \cdot \bar{y}[(n+1) \cdot h]. \tag{40}$$

The coefficients a_n of the series (40) obviously depend only on the value of α and it is evident that for any set constant α these coefficients could be calculated beforehand.

Then the value of h can be determined for every current value of the variable τ, for which the Laplace transformation was made, by the formula

$$h = \frac{1}{\tau} \ln \frac{1}{\alpha}. \tag{41}$$

Thus for a set value of α and a set order of the polynomial approximating the original, we can calculate beforehand the values of the coefficients a_n and after that find the values of the original $y(\tau)$ for every set τ by the formula (40) for which h is determined by (41).

As the calculations showed, for most practical problems one should not strive to have an approximation of the original by a polynomial of a high power.

Sufficient accuracy is assured with $m = 8$. With further increase in m the coefficients a_n increase sharply in their absolute values and the errors in the calculation of the values of $\bar{y}(p)$ for a set p and the errors in determining the coefficients a_n will tell upon the final result.

For practical purposes it is convenient to have the coefficients of the series (40) as integers. It could be achieved by way of the values of α, which are arbitrarily chosen.

Thus for $\alpha = \frac{1}{2}$ we will have

$$y(\tau) \simeq c_m \sum_{n=0}^{m} a_n \cdot h \cdot \bar{y}(p) \tag{42}$$

where

$$h = \ln(2)/\tau; \quad p = (n+1) \cdot h; \quad a_0 = 1; \quad a_2 = -88;$$

$$a_2 = 1804; \quad a_3 = -14872; \quad a_4 = 60346; \quad a_5 = -131560;$$

$$a_6 = 157300; \quad a_7 = -97240; \quad a_8 = 24340.$$

The formula (42) was employed to find the originals of the function by their images.

We will offer some mathematical models of the process of purification of the working medium in a closed volume, which differ by determining the rate of the process stages, the influence of the longitudinal effects and the source functions.

Let us assume the equations (2), (3) are correct but the influence of the longitudinal effect on the dynamics of adsorption is considerable and cannot be accounted for by the correction to the effective coefficient of mass-exchange, according to the expression.

Then instead of (2) we shoule write the material balance equation in the following form:

$$\frac{1}{G}\frac{\partial y}{\partial \tau} = D\frac{\partial^2 y}{\partial \xi^2} - \frac{\partial y}{\partial \xi} - Z\frac{\partial \gamma}{\partial \tau}, \quad Z = \frac{1-\varepsilon}{\varepsilon} \tag{43}$$

and the initial and boundary conditions for the filter are defined thus:

$$\begin{aligned} &y(0, \xi) = \gamma(0, \xi) = 0; \\ &y_0(\tau) = y(0^+, \tau) - D\frac{\partial y(0^+, \tau)}{\partial \xi}; \quad \frac{\partial y(H^-, \tau)}{\partial \xi} = 0. \end{aligned} \tag{44}$$

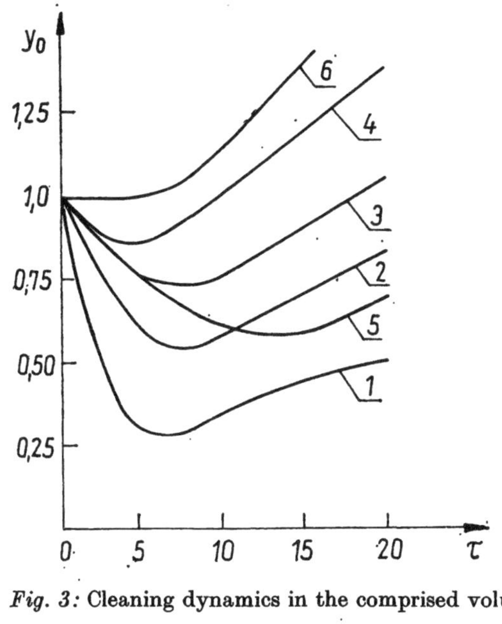

Fig. 3: Cleaning dynamics in the comprised volume.

$D \to 0$; H: *1, 2, 3, 6* – 10; *4* – 5; *5* – 20; K: *1* – 0.415; *2* – 0.208; *3,4,5,6* – 0.138; $K1$: *1,2,3,4,5* – 0.069; *6* – 0.138.

The initial condition for (17) does not change with this. Solving the system (3), (17), (42), (43) by the operator method, we have

$$\bar{y}_0 = \frac{1 + k_1/p}{p + k(1 - F)}, \tag{45}$$

$$F = \frac{1}{M}\left(1 - \frac{Z_2}{Z_1}\right) e^{z_2 \cdot H}, \tag{46}$$

$$M = 1 - k_2 + DZ_1 \cdot k_2 - DZ_2, \tag{47}$$

$$Z_{1,2} = \frac{1 \pm \sqrt{1 + 4D \cdot N}}{2D}, \tag{48}$$

$$k_2 = \frac{Z_2}{Z_1} \exp\left(H(Z_2 - Z_1)\right), \tag{49}$$

$$N = \frac{Zp}{1 + p} + \frac{p}{G}, \tag{50}$$

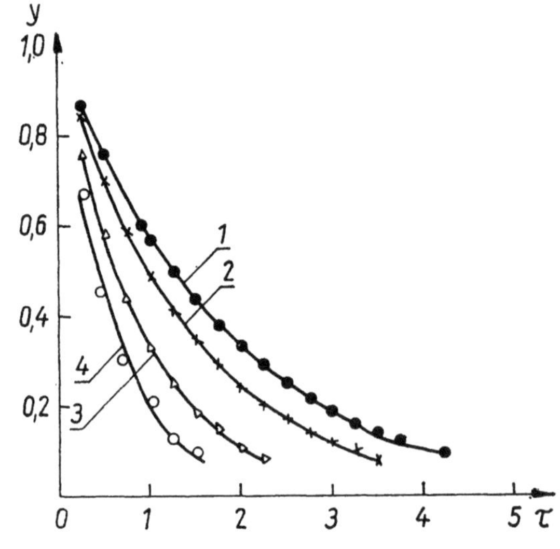

Fig. 4: Cleaning dynamics in the comprised volume.

$H = 10$; $K = 0.415$; $K1 = 0.069$;
D: 1 – 0.05; 2 – 0.1; 3 – 0.15; 4 – 0.25; $G = 10^3$.

It is evident from Fig. 4 that the longitudinal effects lead to a certain lessening of the amplitude of oscillations concentration in the volume W_0.

If the rate of adsorption is determined by a mutual resistance of inner and outer diffusion the expression for determining the image $\bar{y}_0(p)$ remains in the form (45), the auxiliary function F, M, $Z_{1,2}$ and k_2 are determined by the formulas (46)—(49) and

$$N = L + \frac{P}{G} \tag{51}$$

where

$$L = \frac{\sqrt{P \cdot Bi} \operatorname{cth} \sqrt{P \cdot Bi} - 1}{\sqrt{P \cdot Bi} \operatorname{cth} \sqrt{P \cdot Bi} - 1 + Bi/3} \tag{52}$$

for the spherical adsorbent grains with the radius R, $Bi = \frac{\beta_0 R^2}{D_i G}$,

It is not difficult to obtain a similar solution also for the adsorbent grain with different shapes.

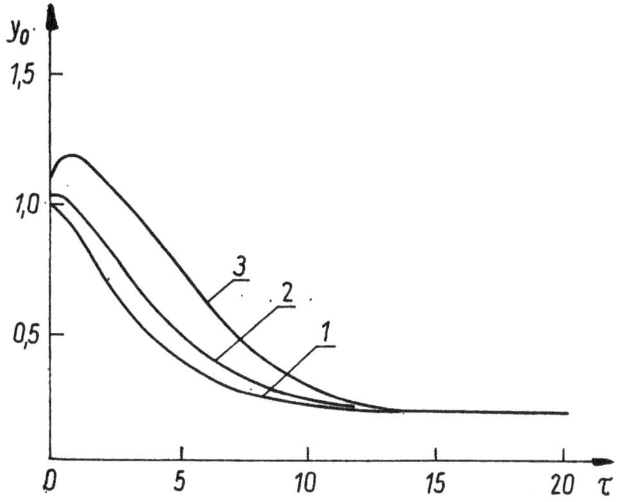

Fig. 5: Cleaning dynamics in the comprised volume. Semidiffusional kinetics.

$H = 10$; $K = 0.415$; $K1 = 0.069$; Bi: *1* — 0.05; *2* — 0.25; *3* — 1.0; $G = 10^3$.

In Fig. 5 some examples of calculations of the process of purification of the working medium for the mixed-diffusional kinetics of adsorption are offered.

A numerical transition from the image of function to its original renders possible solutions which are sufficiently accurate for most practical purposes and can even be used for more complex problems. Also the speed of obtaining the solution does not as a matter of fact depend on the complexity of the mathematical model of the process.

3. The dynamics of the processes of purification of the working medium in a closed volume with a non-linear isotherm

In order to make a qualitative evaluation of the influence of nonlinearity of the isotherm on the change in the concentration in the volume W_0 we will first consider the borderline case of a sharply convex isotherm of adsorption and velocity of the process subject to the equation

$$\frac{\partial a}{\partial t} = \beta_{\text{eff}} \cdot c \cdot \left(1 - \frac{a}{a_\infty}\right), \quad \text{or} \quad \frac{\partial \gamma}{\partial \tau} = y(1-\gamma) \tag{53}$$

where a_∞ —const., $\gamma = \dfrac{a}{a_\infty}$.

If we stipulate that $t \gg \dfrac{x}{v}\,\varepsilon$ from (2) (53) then

$$\frac{\partial^2 \ln y}{\partial \xi\, \partial \tau} + \frac{\partial y}{\partial \xi} = 0, \tag{54}$$

where $\tau = \dfrac{\beta_{\text{eff}} \cdot C_0}{a_\infty} \cdot t;\quad \xi = \dfrac{\beta_{\text{eff}} \cdot x}{v}\,(1 - \varepsilon)$.

By integrating (54) by ξ we obtain

$$\frac{\partial \ln y}{\partial \tau} + y = A(\tau) \tag{55}$$

This expression is correct for $\xi = 0$ as well as for $\xi = H$, then $\xi = 0$

$$\frac{\partial \ln y_0}{\partial \tau} + y_0 = A(\tau) \tag{56}$$

and when $\xi = H$ then

$$\frac{\partial \ln y_H}{\partial \tau} + y_H = A(\tau). \tag{57}$$

Hence $\dfrac{\partial \ln (y_0/y_H)}{\partial \tau} + y_0 - y_H = 0$ and the difference $y_0 - y_H$ in the equation of material balance for the volume W_0 (6) may be replaced by the value for it from the expression obtained above.
Then

$$\frac{\partial \ln (y_0/y_H)}{\partial \tau} - \frac{1}{k}\frac{\partial y_0}{\partial \tau} + \frac{q}{QC_{00}} = 0. \tag{58}$$

Hence, after integrating by τ we get

$$\ln (y_H/y_0) - y_0/k + q \cdot \tau/QC_{00} = B(H),$$
$$B(H) = H - 1/k. \tag{59}$$

For each moment τ

$$y_H = y_0 \cdot F \cdot \exp\left(\frac{k_1}{k}\tau - \frac{y_0}{k}\right),$$

where

$$F = \exp\left(\frac{1}{k} - H\right).$$

Substituting these expressions into (6) we obtain an equation

$$\frac{dy_0}{d\tau} + ky_0\left(1 - \exp\left(-\frac{y_0}{k} + \frac{k_1}{k}\tau\right) + \frac{1}{k} - H\right) - k_1 = 0, \tag{60}$$

for which a solution is easily found by numerical integration.

As the capacity of the adsorbent is limited by the value of a_∞ then if $\tau \to \infty$ the concentration y_H will be equal to y_0. Then from (60) we get

$$1 = \exp\left(\frac{1}{k} - H - \frac{y_0}{k} + \frac{k_1}{k}\tau\right) \tag{61}$$

or

$$1 - kK - y_0 + k_1\tau = 0.$$

Thus for large τ the concentration of impurities in the working medium of a closed volume depends linearily on time

$$y_0 = k_1\tau - (kH - 1).$$

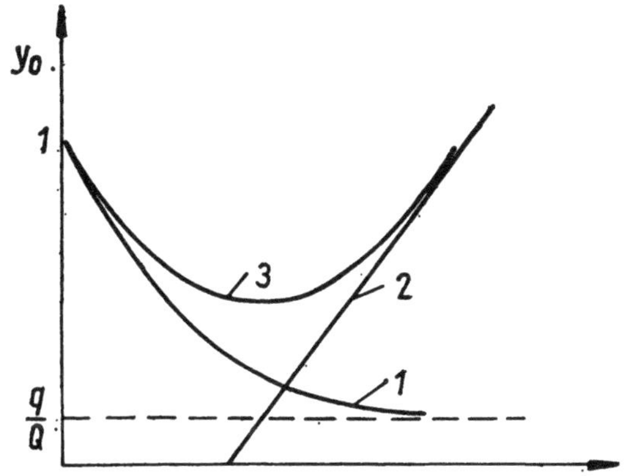

Fig. 6: Cleaning dynamics in the comprised volume. Strong concave isotherm.

1 — $y_0 = y_0(\tau)$, $y_H \to 0$;
2 — $y_0 = y_0(\tau)$, $\gamma \to 1$;
3 — $y_0 = y_0(\tau)$, $\gamma < 1$, $y > 0$.

On the whole, the concentration in the volume will change in the way shown in Fig. 6. The straight line determines the changes of y_0 after the adsorbent filter is saturated. While the concentration beyond the filter y_H is small as compared to y_0, the value of y_0 decreases according to the dependence

$$y_0 \simeq \frac{k_1}{k} + \left(1 - \frac{k_1}{k}\right) \exp(-k\tau). \tag{62}$$

With the rise in the degree of utilization of the adsorptive capacity of the bed, the influence of the concentration beyond the filter y_H becomes more and more considerable and the actual dependence $y_0 = y_0(\tau)$ will correspond to the curve (Fig. 6).

Unlike the processes of purification of the working medium in a closed volume, which take place in a linear region of the isotherm of adsorption, a sharply convex isotherm does not allow for the arising of periodic fluctuations of the concentration y_0.

A case of a sharply convex isotherm is, naturally, the limit to which the process tends if the convexity of the isotherm increases.

The intermediate situations were considered by the authors with reference to the following mathematical model of the process

$$\frac{1}{G}\frac{\partial y}{\partial \tau} = D\frac{\partial^2 y}{\partial \xi^2} - \frac{\partial y}{\partial \xi} - Z\frac{\partial \gamma}{\partial \tau}, \tag{63}$$

$$\frac{\partial \gamma}{\partial \tau} = y - \frac{p\gamma}{1 + p - \gamma}, \tag{64}$$

$$\frac{dy_0}{d\tau} = R \cdot k_1 - k(y_0 - y_H). \tag{65}$$

The initial and boundary conditions

$$y(\xi, 0) = \gamma(\xi, 0) = 0$$

$$y_0(\tau) = y(0^+, \tau) - D\frac{\partial y(0^+, \tau)}{\partial \xi}; \qquad \frac{\partial y(H^-, \tau)}{\partial \xi} = 0.$$

Here

$$y = \frac{C}{C_{00}}; \quad \gamma = \frac{a}{a_0}; \quad y_0 = \frac{C_0}{C_{00}}; \quad y_H = \frac{C_h}{C_{00}}; \quad Z = \frac{1-\varepsilon}{\varepsilon};$$

$$a_0 = \frac{A_\infty, b \cdot C_{00}}{1 + b\,C_{00}}; \quad \xi = \frac{\beta_0 x}{v}\varepsilon; \quad \tau = \frac{\beta_0 t}{G}; \quad G = \frac{a_0}{C_{00}};$$

$$P = \frac{1}{bC_{00}}; \quad k_1 = \frac{q \cdot G}{W_0 \cdot \beta_0 \cdot C_{00}}; \quad k = \frac{Q \cdot G}{W_0 \cdot \beta_0}; \quad R = \exp(-k_H \cdot \tau);$$

$$k_H = f \cdot G/\beta_0; \quad A_\infty, b, f \text{ — constants.}$$

The solution of this problem was obtained by numerical method. We employed the implicit differences scheme of the second order of accuracy over the axis ξ and first order of accuracy over the axis τ.

The examples of the calculations are shown in Fig. 7.

In this case the non-linear character of the isotherm and non-equilibrium nature of the process lead to the lowering of the fluctuations of concentration y_0.

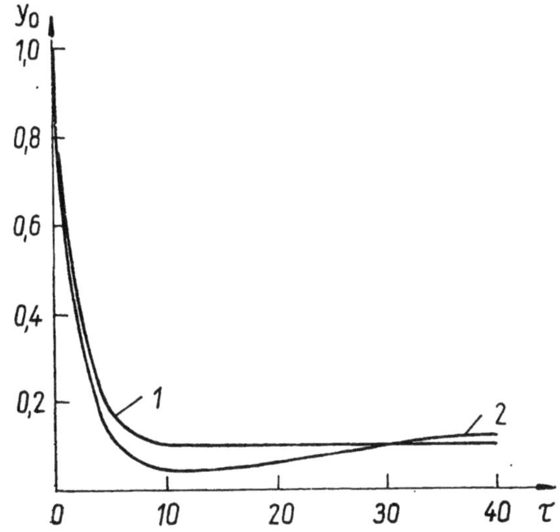

Fig. 7: Cleaning dynamics in the comprised volume.
P: *1* – 0.05; *2* – 0,5; $G = 10^4$; $D = 0.05$; $K = 0.5$;
$Z = 1.5$; $H = 10$; $K_4 = 0$; $R = 1$; *1* – 0.05; *2* – 0.02.

The problem that is of considerable interest is the research into the processes of purification of the working medium in a closed volume from a mixture of various substances. First of all, for many processes it is necessary to purify the working medium in a closed volume from a mixture of harmful products. Secondly, during the process of purification of a mixture of substances the sign reversible two-way process can lead not only to a change in the concentration of a substance but also to changes in the composition of the working medium itself. Moreover, the rate of the introduction of the components of a mixture into the volume W_0 as well as their adsorbability are, in general, different, and the inevitable displacement of some components by the other more adsorbable ones may considerably complicate the laws discussed above of change of the concentrations in a volume.

In case of the purification of a volume W_0 from a mixture of substances, we will have for the process of mixing in a volume W_0 the following system of equations:

$$\frac{dC_{0i}}{dt} = \frac{q_i}{W_0} - \frac{Q}{W_0}(C_{0i} - C_{hi}), \quad i = 1, 2, \dots, n; \tag{66}$$

$$Ch_i = f(C_1, C_2, \dots, C_n; a_1, a_2, \dots, a_n).$$

Let us assume that the rate of adsorption is so high that the process may be considered in equilibrium and the isotherm of sorption for the mixture is described by an equation of LANGMUIR's. Then the process in the adsorbent filter will be in a form of the following system of differential equations.

$$\frac{\partial C_i}{\partial t} = D_i^* \frac{\partial^2 C_i}{dx^2} - v \frac{\partial C_i}{\partial x} - \frac{\partial a_i}{\partial t_n} \tag{67}$$

$$a_i = A_{\infty i} b_i C_i / \left(1 + \sum_{r=1}^{n} b_r c_r\right) \tag{68}$$

Let us introduce into (66)—(68) some non-dimensional variables and parameters

$$y_{0i} = \frac{C_{0i}}{C_{00i}}; \quad y_i = \frac{C_i}{C_{00i}}; \quad y_{hi} = \frac{C_{hi}}{C_{00i}}; \quad P_i = \frac{A_{\infty i}}{b_i C_{00i}};$$

$$Y_i = \frac{y_i}{P_i \left(1 + \sum_{m=1}^{n} \frac{y_m}{P_m} \cdot \frac{A_{\infty m}}{A_{\infty i}}\right)}; \quad \xi = \frac{x}{h}; \quad \tau = \frac{v \cdot t C_{001}}{h \cdot A_{\infty 1}};$$

$$Pe_i = \frac{v \cdot h}{D_i^*}; \quad k = \frac{W_\varphi}{W_0} \cdot \frac{A_{\infty 1}}{C_{001}}; \quad k_i = \frac{q_i}{C_{00i} \cdot Q}; \quad M_i = \frac{A_{\infty i}}{C_{00i}}; \quad \xi = \frac{x}{h}.$$

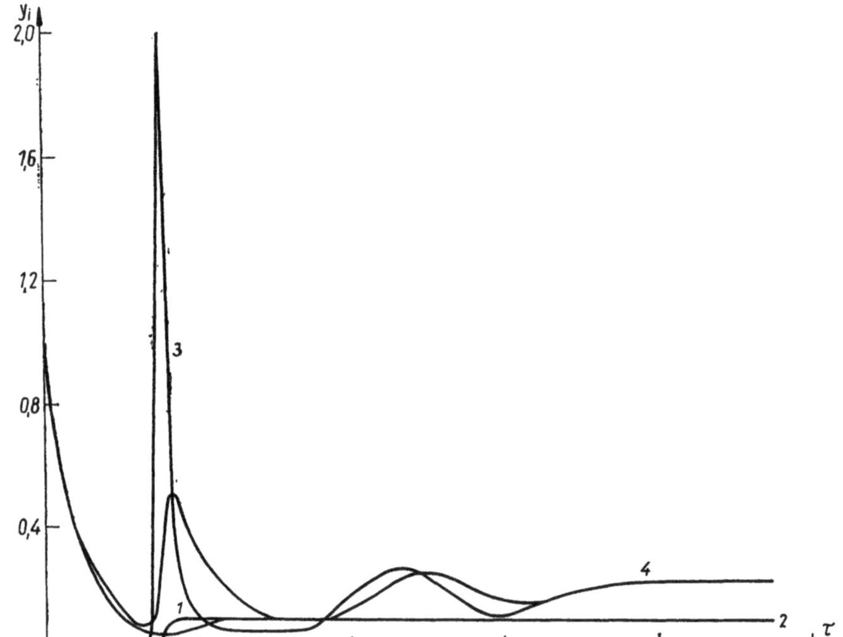

Fig. 8: Sorption dynamics of binary mixture in the comprised volume.
$1 - y_1(1, \tau)$; $2 - y_{01}(\tau)$; $3 - y_2(\tau)$; $4 - y_{02}(\tau)$;
$M_1 = 9.22 \cdot 10^3$; $M_2 = 8.32 \cdot 10^5$; $K_1 = 10^{-4}$; $K_2 = 10^{-3}$;
$K = 1.143$; $Pe_1 = Pe_2 = 50$; $P_1 = 1.34 \cdot 10^{-3}$; $P_2 = 0.252$.

Then we will have

$$\frac{dy_{0i}}{d\tau} = k \cdot k_i - k\,(y_{0i} - y_{hi}) \tag{69}$$

$$\frac{1}{M_1} \cdot \frac{\partial y_i}{\partial \tau} = \frac{1}{Pe_i} \cdot \frac{\partial^2 y_i}{\partial \xi^2} - \frac{\partial y_i}{\partial \xi} - \frac{M_i}{M_1} \cdot \frac{\partial Y_i}{\partial \tau}. \tag{70}$$

We will further take the initial and boundary conditions:

$$y_i(0, \xi) = Y_i(0, \xi) \doteq 0$$

$$y_{0i}(\tau) = y_i(0^+, \tau) - \frac{1}{Pe_i} \cdot \frac{\partial y_i(0^+, \tau)}{\partial \xi}; \quad \frac{\partial y(1^-, \tau)}{\partial \xi} = 0. \tag{71}$$

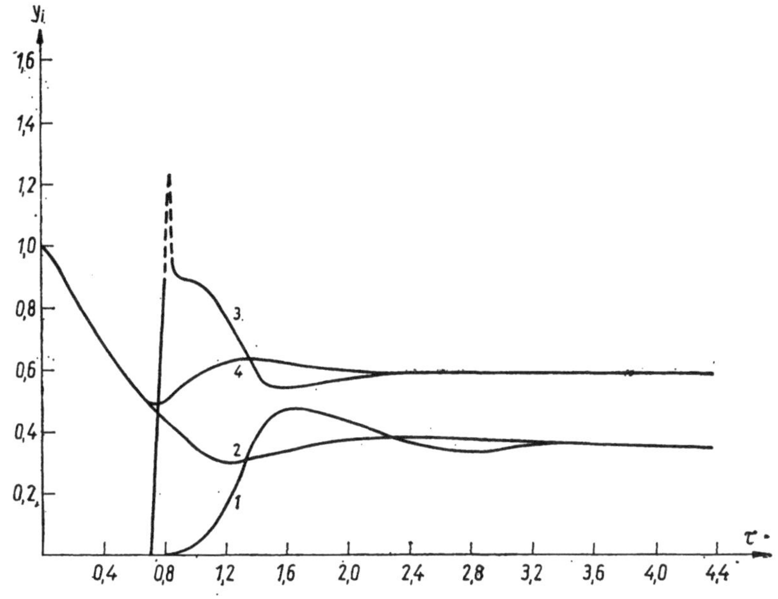

Fig. 9: Sorption dynamics of binary mixture in the comprised volume.
1 — $y_1(1, \tau)$; 2 — $y_{01}(\tau)$; 3 — $y_2(1, \tau)$; 4 — $y_{02}(\tau)$;
$M_1 = 9.22 \cdot 10^3$; $M_2 = 8.32 \cdot 10^3$; $K_1 = 10^{-4}$;
$P_1 = 1.34 \cdot 10^{-3}$; $P_2 = 2.52 \cdot 10^{-3}$.

We used the implicit scheme of numerical solution for this system. The solution was obtained on a digital computer.

Some of the results, shown in Figs. 8, 9, indicate that in the case there also take place fluctuations of the concentration of the sorbate component in a volume W_0. The relaxation fluctuation of the concentration are also further complicated by the process of displacement of less adsorbable components of the mixture.

As the displacement of a component from the adsorbed state inevitably leads to an increase in concentration of this component in the flow leaving the sorbent layer, this circumstance leads to an increase in the amplitude of the fluctuations.

Calculation showed that in some cases the concentration of one (or more) components of the mixture in some separate moments can be considerably higher than the initial value of this concentration.

This fact is obviously important for the development and analysis of the functioning of the purifying processes for the medium in closed volumes, such as e.g. air purification or haemosorption.

4. Summary

When there is a process of purification by adsorption of a medium circulating in a closed volume, there may occur relaxation fluctuations of concentration in this volume; the amplitude of the fluctuations, which are due to sign reversible two-way process, is the higher, the higher the rate of adsorption is and the more the sorption isotherm approaches a linear isotherm.

The displacement effects occuring during the mixture adsorption can increase the amplitude of fluctuations of the components concentration in the volume and also make the values of concentration exceed the initial values.

5. References

[1] Wicke, E., Koll. Z. **86**, 1, 67, 295 (1939); **90**, 156 (1940); **93** 123, 159 (1940).
[2] Wilson, N., J. Am. Chem. Soc. **62**, 1563 (1940).

Mathematical modelling of cyclic processes

(A. K. Akulov, E. A. Ustinov, V. L. Kolin, I. N. Taganov)

Zusammenfassung:

Es werden mathematische Modelle für Prozesse, die aus den hintereinander liegenden Stufen Adsorption, Desorption Abkühlung beliebiger Zeitdauer bestehen, aufgestellt. Dazu werden numerische Lösungsmethoden für die EDVA und Beispiele angeführt.

An Hand der berechneten Lösungen werden qualitative und quantitative Gesetzmäßigkeiten untersucht. Es wird gezeigt, daß sich bei der Konzentration und der Temperatur in den einzelnen Phasen in jedem Stadium des nachfolgenden Zyklus die entsprechenden Verteilungen der gleichen Stufen des vorhergehenden Zyklus wiederholen. Es werden Rechnerergebnisse von Adsorptions-Desorptionsprozessen angeführt und mit experimentellen Daten verglichen.

In various fields of chemical technology the cyclical method began to be widely employed lately, which allows to considerably intensify such processes as extraction, rectifying, filtration, etc. [1]. There is also a group of technological processes which are traditionally carried out in a cyclic way. Among these are first of all adsorption and ion-exchange processes, for which the technological cycle consists of consecutively recurrent stages of sorption-desorption, which are supplemented when necessary by such stages as drying, cooling or washing.

The standard of theoretical works on the cyclic processes nowadays obviously falls short of their practical importance [2, 3]. In this paper the authors introduce a single method of approach to the mathematical modelling of cyclic processes employing the theory of operators [4].

Let us consider some chemical-technological object which at any moment of time may be identified by a K-dimensional vector of state w, of which the elements are independent parameters of a process, including the functions of distribution of concentrations, temperatures, etc. Let us assume that the operating influence is carried out in n stages successively proceeding in time and comprising a single cycle.

We shall consider the object state not at every moment of time but at a certain characteristic point, for instance in the middle or at the end of the stage. The object state at such point of jth stage of rth cycle is estimated by the vector $\boldsymbol{w}_j^{(r)}$. The multitude of vectors $\boldsymbol{w}_j^{(r)}$ together with the operations of adding and substraction properly introduced comprise a linear space of states Ω:

$$\boldsymbol{w}_j^{(r)} \in \Omega; \quad j = \overline{1, n}; \quad r = 1, 2, \ldots$$

Let us introduce into the space Ω the operators $\mathscr{A}_j$ according to the rule:

$$\boldsymbol{w}_j^{(r)} = \begin{cases} \mathcal{A}_j \boldsymbol{w}_{j-1}^{(r)}, & j = \overline{2, n}; \\ \mathcal{A}_1 \boldsymbol{w}_n^{(r-1)}, & j = 1; \end{cases} \quad r = 1, 2. \ldots \tag{1}$$

and name $\mathscr{A}_j$ the operator for the jth stage.

Equation (1) is the recurrent procedure for the calculation of any vector $w_j^{(r)}$ if the initial state of the object characterized by the vector $w^{(0)}$ and the form of the operators $\mathcal{A}_j$ are stated. If we confine ourselves to the consideration only of the first r cycles then (1) may be regarded as a system of $n \times r$ operator equations.

With regard to (1) we will write the expression for the definition of $w_j^{(r)}$ in a distinct form:

$$w_j^{(r)} = \mathcal{A}_{j1}\mathcal{A}_{n1}^{r-1}w^{(0)} \tag{2}$$

where

$$\mathcal{A}_{kl} = \mathcal{A}_k\mathcal{A}_{k-1} \dots \mathcal{A}_{l+1}\mathcal{A}_l$$

$$\mathcal{A}_{kl}^r = \underbrace{\mathcal{A}_{kl}\mathcal{A}_{kl} \dots \mathcal{A}_{kl}}_{r}.$$

The recurrent procedure (1) and the equation (2) can be employed for the investigation of a non-stationary regime of the cyclic process.

Let us consider the conditions for which

$$w_j^{(r)} = w_j^{(r+1)} = w_j, \quad j = \overline{1, n}; \quad r = 1, 5, \dots \tag{3}$$

In the future we will call this condition quasi-steady-state or stationary-cyclical regime. We shall write down an operator equation establishing the connection between the vectors $w_j^{(r)}$ and $w_j^{(r+1)}$ employing (1) for this:

$$w_j^{(r+1)} = \mathcal{A}_{j1}\mathcal{A}_{n(j+1)}w_j^{(r)}, \quad j = \overline{1, n}; \quad r = 1, 2, \dots \tag{4}$$

with regard to (3) from (4) we will obtain n independent operator equations for the quasisteady-state regime to apply further for determination of n vectors of state.

$$w_j = \mathcal{A}_{j1}\mathcal{A}_{n(j+1)}w_j, \quad j = \overline{1, n}. \tag{5}$$

As $\mathcal{A} = \mathcal{A}_{n(j+1)}\mathcal{A}_{j1}$, then after some identical transpositions, we will rewrite (5) as

$$w_j = \mathcal{A}_{j1}\mathcal{A}\mathcal{A}_{j1}^{-1}w_j, \quad f = \overline{1, n} \tag{6}$$

where $\mathcal{A}_{j1}^{-1}$ — the operator reversed to $\mathcal{A}_{j1}$.

The analytical solution for the operator equations (5) or (6) can be obtained for certain specified cases.

Let us consider for example a two-stage cyclic process in the EUCLID space. Let us further assume that the stage operators are set by the matrix equations:

$$\begin{aligned} w_1^{(r)} &= A_1 w_2^{(r-1)} + B_1, \\ w_2^{(r)} &= A_2 w_1^{(r)} + B_2, \end{aligned} \qquad r = 1, 2, \dots$$

where A_1, A_2 — square matrices of k order; B_1, B_2 — matrices-columns of k dimensions. For the stationary-cyclical regime according to (3) the following system of equations will be correct:

$$\boldsymbol{w}_1 = A_2\boldsymbol{w}_2 + B_1,$$
$$\boldsymbol{w}_2 = A_2\boldsymbol{w}_1 + B_2. \qquad (7)$$

From the system (7) we obtain two independent matrix equations in respect to the vectors of state $\boldsymbol{w}_1$ and $\boldsymbol{w}_2$

$$(E - A_1A_2)\,\boldsymbol{w}_2 = A_2B_1 + B_2 \qquad (8)$$
$$(E - A_1A_2)\,\boldsymbol{w}_1 = A_1B_2 + B_1 \qquad (9)$$

where E is a unitary matrix of k order.

The solutions for equations (8), (9) may be written in the following form

$$\boldsymbol{w}_2 = (E - A_2A_1)^{-1}\,(A_2B_1 + B_2),$$
$$\boldsymbol{w}_1 = (E - A_1A_2)^{-1}\,(A_1B_2 + B_1).$$

For the general case the solution of the operator equation (6) can be obtained by the application of the method of successive approximations. As a zero approximation $\boldsymbol{w}_{j0}$ we will take the vector of state for the first cycle:

$$\boldsymbol{w}_{j0} = \boldsymbol{w}_j^{(1)} = \mathcal{A}_{j1}\boldsymbol{w}^{(0)}.$$

The first approximation can be determined from (6) in the following way:

$$\boldsymbol{w}_{j1} = \mathcal{A}_{j1}\mathcal{A}\mathcal{A}_{j1}^{-1}\boldsymbol{w}_{j0} = \mathcal{A}_{j1}\mathcal{A}\mathcal{A}_{j1}^{-1}\mathcal{A}_{j1}\boldsymbol{w}^{(0)} = \mathcal{A}_{j1}\mathcal{A}\boldsymbol{w}^{(0)} = \boldsymbol{w}_j^{(2)}.$$

The 1st approximation coincides with the vector of state for the 2nd cycle. And by analogy we obtain

$$\boldsymbol{w}_{j2} = \mathcal{A}_{j1}\mathcal{A}\mathcal{A}_{j1}^{-1}\boldsymbol{w}_j^{(2)} = \mathcal{A}_{j1}\mathcal{A}\mathcal{A}_{j1}^{-1}\mathcal{A}_{j1}\mathcal{A}\boldsymbol{w}^{(0)} = \mathcal{A}_{j1}\mathcal{A}^2\boldsymbol{w}^{(0)} = \boldsymbol{w}_j^{(3)}.$$

We generalize the results by applying the method of mathematical induction and obtain the conclusion that the rth approximation is the vector of state for the $(r + 1)^{\text{th}}$ cycle

$$\boldsymbol{w}_{jr} = \boldsymbol{w}_j^{(r+1)}.$$

Hence the method of successive approximation when applied to the solution of operator equation (6) coincides with the process of successive calculations of the vectors of state from the operator equation for cyclic process (2) carried out up to the achievement of the accuracy set beforehand of the establishment of the quasisteady-state regime. As the choice of the initial approximation $\boldsymbol{w}^{(0)}$ has been made arbitrarily one can draw the conclusion that if the model operators $\mathcal{A}_1, \ldots, \mathcal{A}_n$ are adequate to the real ones the calculations based on the equation (2) will regardless of the initial approximation taken lead to the obtainment of the vectors of state for the stationary-cyclical regmie.

As an example we will consider a two-stage adsorption-desorption process. Then we will make the following assumptions: the kinetics of the process is limited by the outer diffusion; the process is isothermal; the flow velocity, entrance concentration and kinetic coefficient depend only on the half-cycles number; isotherm of sorption is described by LANGMUIR's equation stretched for the case of mixtures adsorption. Also the dynamics of countercurrent adsorption of an m-component mixture for the even ($z = 1$) and uneven ($z = 0$)-numbered half-cycles can be described by the following mathematical model

$$\frac{1}{\Gamma_1}\frac{\partial y_i}{\partial \tau} + \delta_i m \frac{\partial \gamma_i}{\partial \tau} + (-k_v k_t)^z \frac{\partial y_i}{\partial \xi} = k_t^z D \frac{\partial^2 y_i}{\partial \xi^2},$$

$$\frac{\partial \gamma_i}{\partial \tau} = (k_{\beta_i} k_t)^z \frac{r_t}{\delta_i}(y_i - y_i^*), \quad i = \overline{1, M}$$

$$y_i^* = \gamma_i \cdot \frac{\left(1 + \sum\limits_{s=1}^{M} \frac{y_s^*}{p_s}\right)}{\left(1 + \sum\limits^{M} \frac{1}{p_s}\right)}, \quad z = \overline{0, 1}. \tag{10}$$

$$y_i(\xi, 0) = y_{ni}(\xi), \quad \gamma_i(\xi, 0) = \gamma_{ni}(\xi) \cdot$$

$$(-k_v)^z [k_{y_i}^z - y_i(z \cdot L, \tau)]' + D \frac{\partial y_i}{\partial \xi}(z \cdot L, \tau) = 0, \quad \frac{\partial y_i}{\partial \xi}\big(L \cdot (1 - z), \tau\big) = 0.$$

In order to exclude the variables y_i^*, $i = \overline{1, M}$ we will divide both parts of the 3rd equation in the system (10) by the parameter p_i and then add up all M equations. Then we get

$$\sum_{s=1}^{M} \frac{\gamma_s}{p_s} = \left(1 + \sum_{s=1}^{M} \frac{1 - \gamma_s}{p_s}\right) \cdot \sum_{s=1}^{M} \frac{y_s}{p_s}. \tag{11}$$

Applying (11) to the exclusion of the variables y_i^* we get the system:

$$\frac{1}{\Gamma_1}\frac{\partial y_i}{\partial \tau} + \delta_i m \frac{\partial \gamma_i}{\partial \tau} + (-k_v k_t)^z \frac{\partial y_i}{\partial \xi} = k_t^z D \frac{\partial^2 y_i}{\partial \xi^2}$$

$$\frac{\partial \gamma_i}{\partial \tau} = (k_{\beta_i} k_t)^z \frac{r_i}{\delta_i}\left(y_i - \frac{\gamma_i}{1 + \sum\limits_{s=1}^{M} \frac{1 - \gamma_s}{p_s}}\right), \quad i = \overline{1, M}$$

$$y_i(\xi, 0) = y_{ni}(\xi), \quad \gamma_i(\xi, 0) = \gamma_{ni}(\xi), \quad z = \overline{0, 1}; \tag{12}$$

$$(-k_v)^z \big(k_{y_i}^z - y_i(z \cdot L, \tau)\big) + D \frac{\partial y_i}{\partial \xi}(z \cdot L, \tau) = 0,$$

$$\frac{\partial y_i}{\partial \xi}\big(L(1 - z), \tau\big) = 0.$$

Let us choose the space of states as a direct product of $2 \times M$ functional spaces $C^2[0, L]$:

$$\Omega = \underbrace{C^2{}_{[0,L]} \times C^2{}_{[0,L]} \times \dots \times C^2{}_{[0,L]}}_{2 \times M}$$

where $C^2[0, L]$ is the space of twice continously-differentiated functions over the interval $[0, L]$. As an element of the space Ω there is a vector-function $\{y_i, \gamma_i\}_{i=\overline{1,M}}$ of $2 \times M$ dimension, of which the components are the functions of distribution of concentrations of M — components of the mixture in the gas and the sorbent over the length of the adsorbent layer. The nonlinear system of differential equations in partial derivatives with initial data and boundary conditions of a DANKWERTS type (12) gives the operators for even- and uneven-numbered half-cycles $\mathcal{A}_z$. The recurrent procedure for this instance consists of successive solutions of the system (12), beginning from some initial aprroximation and going on until the quasisteady-state regime is established with the required accuracy. Also the initial condition for each half-cycle is determined by the vector of state obtained at the end of the previous half-cycle. For the solution of the system (12) the method of finite differences was employed. The rectangular area $\sigma = [0, L] \times [0, \tau_0]$ where we sought the solution for the system (12) we replace by the net area $\sigma_{\xi\tau} = \sigma_\xi \times \sigma_\tau$ where

$$\sigma_\xi = \left\{\xi_j = j \cdot \Delta\xi,\ j = \overline{0, N}\right\}, \quad \Delta\xi = L/N$$

$$\sigma_\tau = \left\{\tau_k = k \cdot \Delta\tau,\ k = \overline{0, k_0}\right\}, \quad \Delta\tau = \tau_0/k_0.$$

It is known [5] that to apply distinct schemes to the solving of a non-linear parabolic equation is not expedient. The condition of stability requires too small time-paces. Over a six-point scheme there was based an implicit differences scheme of a second order of accuracy along the axis ξ and of a first order of accuracy along the axis τ.

The non-linear member

$$\varphi(\gamma_1, \ldots, \gamma_M) = \frac{1}{1 + \sum\limits_{s=1}^{M} \dfrac{1 - \gamma_s}{p_s}}$$

in the second equation of the system (12) has been calculated at the previous time-layer. This last instance made it possible to reduce the problem to the solving at each time-layer of M independent algebraic systems. For the solution of the systems a method of trial runs was used. The establishment of the quasi-steady-state regime was ascertained by whether the performance has been according to the equation of material balance with the accuracy of 0.1% set beforehand. This equation for the ith component may be written in the following form:

$$1 + k_v k_t k_{y_i} = \frac{1}{\tau_0} \int\limits_0^{\tau_0} y_i^{(0)}(L, \tau)\, d\tau + \frac{k_v}{\tau_0} \int\limits_0^{\tau_0 k_t} y_i^{(1)}(0, \tau)\, d\tau$$

The above described algorithm was put into effect asa computer programme in Algol-60. In Fig. 1 there are given, as an example, the functions of distribution of the concentrations in the mobile phase (gas, liquid) at the end of an uneven half-cycle for the zero initial filling of the sorbent layer. The calculations showed that the number of cycles, before the stationary-cyclic regime is established, may be counted by hundreds if the, initial approximation is used. The highest number of cycles were required when the, values of parameters τ_0/L, p were in the ranges $0 < \tau_0/L < 0.1$, $0 < p < 0.5$.

Then in order to obtain the distribution of concentrations for the quasi-steady-state regime it is essential to use reasonable initial approximations.

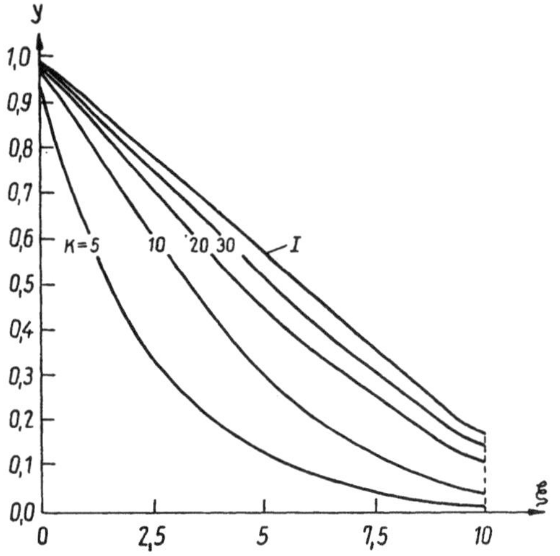

Fig. 1: The development of a cyclic process.
The end of adsorption. The number of cycle — k;
stationary-cyclical distribution — I;
$p \to \infty$; $D = 0.25$; $L = 10$; $\tau_0 = 1$; $\varepsilon = 0.4$; $k_v = k_t = k_\beta = 1$; $k_y = 0$; $\Gamma = 10^4$.

Let us further consider an adsorption-desorption cyclic process for one component and a linear isotherm. If we ignore the effect of longitudinal diffusion and assume that HENRY's coefficient Γ is sufficiently large then the initial system of equations can be written as:

$$\begin{aligned} \frac{\partial y}{\partial \xi} &= m\left(-\frac{k_\beta}{k_v}\right)^z \cdot \left(\frac{\gamma}{k_\Gamma{}^z} - y\right), \quad y(z \cdot L, \tau) = k_y{}^z; \\ \frac{\partial \gamma}{\partial \tau} &= (k_t k_\beta)^z \cdot \left(y - \frac{\gamma}{k_\Gamma{}^z}\right), \quad \gamma(\xi, 0) = \gamma_n(\xi); \end{aligned} \qquad r = \overline{0,1}. \quad (13)$$

Let us assume that the half-cycle time is small. Then we approximate the function of changes in time of the concentration in the sorbent for the ith cycle by a polynomial of 1st order:

$$\gamma^{(i)}(\xi, \tau) = (-1)^z \left[\gamma_0{}^{(i)} - \gamma_1{}^{(i+z-1)}\right] \frac{\tau}{\tau_0} + (1 - z)\, \gamma_1{}^{(i-1)} + z\gamma_0{}^{(i)} \qquad (14)$$

where $\gamma_0{}^{(i)}(\xi)$ and $\gamma_1{}^{(i)}(\xi)$ are concentrations at the end of uneven and even half-cycles. We will write down the system (13) with regard to (14) for the ith

cycle at the moments of time which correspond to the middle of even and uneven half-cycles.

$$\frac{dy_z^{(i)}}{d\xi} = m\left(-\frac{k_\beta}{k_v}\right)^z \left\{(-1)^z \frac{[\gamma_0^{(i)} - \gamma_1^{(i+z-1)}]}{2k_\Gamma^z} - y_z^{(i)}\right\}$$
$$(-1)^z \frac{1}{\tau_0}[\gamma_0^{(i)} - \gamma_1^{(i+z-1)}] = (k_\beta k_t)^z \left\{y_z^{(i)} - \frac{1}{2k_\Gamma^z}[\gamma_0^{(i)} - \gamma_1^{(i+j-1)}]\right\} \tag{15}$$

where $y_z^{(i)}(\xi)$ — mean concentration in the mobile phase during z^{th} half-cycle of ith cycle.

The first $4 \times n$ equation (15) which corresponds to the first n cycles can be regarded as a system of $2 \times n$ ordinary differential and $2 \times n$ algebraic equations relatively to the variables $\gamma_z^{(i)}$, $y_z^{(i)}$, $i = \overline{1, n}$; $z = \overline{0, 1}$. The system (15) can be reduced to a system of ordinary differential equations by exluding the variables $\gamma_z^{(i)}$:

$$\frac{(2+\tau_0)}{2}\frac{dy_0^{(i)}}{d\xi} + \frac{k_\Gamma}{2}(2\varkappa - \alpha\tau_0)\frac{dy_1^{(i-1)}}{d\xi} = (k_\Gamma y_1^{(i-1)} - y_0^{(i)})\,m, \quad i = \overline{2, n}$$
$$\frac{(2-\tau_0)}{2}\frac{dy_0^{(i)}}{d\xi} + \frac{k_\Gamma}{2}(2\varkappa + \alpha\tau_0)\frac{dy_1^{(i)}}{d\xi} = (k_\Gamma y_1^{(i)} - y_0^{(i)})\,m, \quad i = \overline{1, n} \tag{16}$$
$$\frac{(2+\tau_0)}{2}\frac{dy_0^{(1)}}{d\xi} = m(\gamma_n - y_0^{(1)})$$
$$y_0^{(i)}(0) = 1, \quad y_1^{(i)}(L) = k_y.$$

For the quasisteady-state regime the distributions of concentrations for the $(i+1)^{\text{th}}$ cycle coincide with those for the ith cycle:

$$y_z^{(i+1)}(\xi) = y_z^{(i)}(\xi), \quad z = \overline{0, 1}. \tag{17}$$

The differential equation describing the stationary-cyclic distribution of concentrations in the mobile phase can be obtained from (16) with regard to (17):

$$\frac{d^2y_z}{d\xi_1} + \lambda\frac{dy_z}{d\xi} = 0 \tag{18}$$
$$(-\alpha)^{1-z}\{k_\Gamma^{1-2z}k_y^{1-z} - y_z[L(1-z)]\}$$
$$+ (\alpha + \varkappa)\frac{dy_z}{d\xi}[L(1-z)] = 0, \quad y_z(z \cdot L) = k_y^z.$$

There is an analytical solution to the equation (18) which can be obtained in the following form

$$y_z(\xi) = \begin{cases} \dfrac{\alpha k_\Gamma k_y - \exp(-\lambda L) + \alpha^{1-z}[(1 - k_\Gamma k_y)\exp(-\lambda\xi)}{k_\Gamma^z[\alpha - \exp(-(\lambda L)]}, & \alpha \neq 1; \\ k_y^z - \dfrac{(1-k_\Gamma k_y)\,m}{k_\Gamma^z(1+\varkappa+Lm)}(\xi - Lz), & \alpha = 1 \end{cases} \tag{19}$$

The function of distribution of the mean concentration in the sorbent we will obtain by employing the kinetics equation:

$$\gamma_m(\xi) = \frac{\alpha}{(\alpha + \varkappa)} \left(k_\Gamma y_1 + \frac{\varkappa}{\alpha} y_0 \right). \tag{20}$$

The functions of changes in concentrations during the quasisteady-state regime cycle in time can be expressed through the solutions obtained:

$$\begin{aligned} \gamma(\xi, \tau) &= \gamma_m(\xi) + (-1)^z \frac{dy_1}{d\xi} \left(\frac{\tau_0}{2} - \tau \right), \\ y(\xi, \tau) &= y_z(\xi) + \frac{1}{(-k_\Gamma)^z} \frac{dy_1}{d\xi} \left(\frac{\tau_0}{2} - \tau \right), \quad z = \overline{0, 1}. \end{aligned} \tag{21}$$

The analytical solutions (19—21) were compared to the results of the numerical integration of the system (5) when $p \to \infty$. The satisfactory coinciding of the values of concentrations at the calculated points were observed on a wide range of changes of the half-cycle time τ_0.
This recommends the use of the solutions (19—21) for engineering calculations of the isothermal adsorption-desorption process having the linear isotherm.

The numerical experiments showed that in case of non-linear isotherms ($p < \infty$) the distribution of concentrations in the gaseous phase for the low values of τ_0 can also be approximately described by the solution (19) and the concentration in the sorbent by the following correlation:

$$\gamma_{m_p} = \frac{\gamma_m(\xi)\,(p + 1)}{p - \gamma_m(\xi)}. \tag{22}$$

The correlation (22) is obtained by substituting into the equation for Langmuir's isotherm the concentration $\gamma_m(\xi)$. Then the maximum values of the half-cycle time, for which the correlations (19—22) are correct, decrease with the parameter p.

Let us further consider another borderline case satisfying the following provisions: the velocities of the adsorption and desorption processes are so high that these processes may be regarded as equilibrium; the influence of longitudinal diffusion is too small as to be safely ignored; in the whole flow there is only one adsorbable component; the processes of adsorption and desorption are isothermal. The initial system of equations can be written as:

$$\begin{aligned} &\frac{d\gamma}{d\eta} + (-k_v k_i)^z \frac{dy}{d\zeta} = 0, \quad y(z \cdot H, \eta) = k_y^z, \\ &y = \frac{p\gamma}{k_b^z(p + 1 - \gamma)}, \quad \gamma(\zeta, 0) = \gamma_H(\zeta), \quad z = \overline{0, 1}. \end{aligned} \tag{23}$$

From the system (23) expressions for the velocities of movement along the adsorbent layer of the point of constant concentrations y are easily obtained:

$$u = \varphi^z \frac{(p + k_b^z y)^2}{p(p + 1)}, \quad z = \overline{0, 1}. \tag{24}$$

If as is usual for the real processes $p > 0$ then during adsorption there will always form a detached front of the concentration y moving along the layer at the rate $u = 1$. During the desorption the front of the concentrations will be continuously washing away through the whole half-cycle. Then the point of minimum concentration k_y will move at the rate $\varphi \frac{(p + k_y k_b)^2}{p(p + 1)}$ while the point of maximum concentration $\gamma = \frac{1}{k_b}$ in equilibrium to $\gamma = 1$ will move along the layer at the rate $\varphi \frac{(p + 1)}{p}$. The ratio of the width of the front to the length of the path of the point of minimum concentration on the condition that at the initial moment the front was of a detached nature remains the same for any moment of time for adsorption as well as for desorption. The ratio is equal to

$$\frac{(p + 1)^2 - (p + k_y k_b)^2}{(u + k_y k_b)} .$$

If the initial concentration of the component in the desorbing flow is $k_y = 0$ the ratio of the width of the concentration front to the length of the path for the point at which $y = 0$ depends neither on the correlation of velocities of the direct- and counter-flows k_v nor on the correlation of the isotherms constants k_b. Hence it is obvious that the process of adsorption goes on until the bed is completely satiated. At the moment of the emergence of the minimum concentration $k_y k_b$ the detached front of concentrations has already been formed and the concentration y beyond the bed leaps upwards to 1. Indeed, as the ratio of the width of the concentrations front to the co-ordinate of its lower border is constant at the point $\xi = 0$ the width of the front is equal to zero. It is also correct for the adsorption when $\xi = H$ and the front should be of a detached type. This formulated rule should be all the more correct when during desorption at the end of the half-cycle the area of high concentration emerges from the bed. Then during the subsequent adsorption the concentration at the pont $\xi = 0$ will rise to 1 and at the moment the adsorption front leaves the bed this front will be of a detached type.

Thus if during adsorption the bed is satiated up to the uniform and equilibrium to 1 fullness then the form of the emergence curves for the adsorption as well as for desorption do not depend on the initial filling of the bed during the adsorption stage. If the adsorption is incomplete then during the subsequent adsorption at the moment η the concentration of the sorbate beyond the bed is higher than at the same moment of process in case of the bed being originally free of the sorbate. It follows from the equations of material balance that for the stationary regime the total quantity of the substance over the cycle introduced into the bed is equal to the quantity of substance emerged from the bed during adsorption and desorption.

If the rate of movement of the maximum concentration in the front during desorption is less than the rate of movement of the detached front during adsorption and the complete satiation of the bed originally free of sorbate is not achieved during one half-cycle the desorption will always be incomplete and through the half-cycle of every η_0 duration conditions will be formed which will allow a part of the substance entering the bed during the adsorption stage to emerge from the bed during this same stage. If we assume that the entry concentration of the component aimed at in the flow during desorption k_y is equal to zero and the rate of movement of the point of the desorption front where $k_y = 0$ is higher than the velocity of the detached front of concentration $\left(\frac{\varphi \cdot p}{p+1} > 1\right)$ then for each cycle the desorption will be complete and the dynamics of the process will not differ from the dynamics of known frontal processes. Also if $\eta_0 < H$ then the concentration beyond the layer at the moment when the adsorption stage ends is equal to zero.

We write the above-considered qualitative regularities for a cyclic process with convex isotherm ($p > 0$) for the stationary-cyclic conditions as the following quantitative expressions:

a) $\nu < \frac{p}{p+1}, \quad \varphi < 1$

$$y(z \cdot H, \eta) = \begin{cases} 1/k_b, & z = 1; \\ k_y k_b, & z = 0, \quad \eta < \varphi\eta_0 \\ 1, & z = 0, \quad \varphi\eta_0 < \eta < \eta_0; \end{cases}$$

$$y(z \cdot H) = 1 - z + \frac{1}{k_b}\,[-k_v k_t (1 - k_y k_b)]^{1-z}$$

b) $\frac{p}{p+1} < \nu < \frac{p(p+1)}{(p + k_y k_b)^2}, \quad \mu > \frac{1 - (\sqrt{1+p} - \sqrt{p\nu})^2 - \nu k_y k_b}{1 - k_y k_b}$

$$y(z \cdot H, \eta) = \begin{cases} 1/k_b, \quad z = 1, \quad \eta < \dfrac{H}{\varphi}\,\dfrac{p}{(p+1)} \\[2ex] \sqrt{\dfrac{p(p+1)\,H}{k_v k_t k_b \eta}} - \dfrac{p}{k_b}, \quad z = 1, \quad \dfrac{H}{\varphi}\,\dfrac{p}{p+1} < \eta < \eta_0 \\[2ex] k_y k_b, \quad z = 0, \quad \eta < \dfrac{H \cdot [\sqrt{(p+1)\,H} - \sqrt{p\varphi\eta_0}]^2 - k_v k_t k_y \eta_0}{1 - k_y k_b} \\[2ex] 1, \quad z = 0, \quad \eta_0 > \eta > \dfrac{H\,[\sqrt{(p+1)\,H} - \sqrt{p\varphi\eta_0}]^2 - k_v k_t k_y \eta_0}{1 - k_y k_b} \end{cases}$$

$$y(z \cdot H) = 1 - z + (-1)^z \left[-\frac{1}{\mu (k_v k_t)^z} + \left(\sqrt{\frac{p+1}{\mu (k_t k_t)^z}} - \sqrt{\frac{p (k_v k_t)^{1-z}}{k_b}} \right)^2 \right] + (1 - z)\, k_v k_t k_y$$

c) $\nu > \frac{p(p+1)}{(p+k_y k_b)^2}, \quad \mu > \frac{p}{(p+k_y k_b)};$

$$y(z\cdot H, \eta) = \begin{cases} 1/k_b, & z = 1, \quad \eta < \frac{Hp}{\varphi(p+1)}; \\ \sqrt{\frac{p(p+1)\,H}{k_v k_t k_B \eta_0}} + \frac{p}{k_b}, & z = 1, \quad \frac{Hp}{\varphi(p+1)} < \eta < \frac{H(p+1)\,p}{\varphi(p+k_y k_b)^2} \\ k_y, & z = 1, \quad \frac{Hp(p+1)}{\varphi(p+k_y k_b)^2} < \eta < \eta_0 \\ k_y k_b, & z = 0, \quad \eta < \frac{Hp}{p+k_y k_b} \\ 1, & z = 0, \quad \frac{Hp}{p+k_y k_b} < \eta < \eta_0 \end{cases}$$

$$y_z(z\cdot H) = k_y{}^z - (-1)^z \frac{1}{\mu(k_v k_z)^z}\cdot\frac{p(1-k_y k_b)}{(p+k_y k_b)}$$

d) $\mu < \frac{p\eta_0}{\varphi B(p+1)}, \quad 1 < \varphi < \frac{p+1}{p+k_y k_b}$

$$y(z\cdot H, \eta) = \begin{cases} 1/k_b, & z = 1, \quad \eta < B \\ \frac{(p+1)}{k_b}\sqrt{\frac{B}{p\eta}} - \frac{p}{k_b}, & z = 1, \quad B < \eta < \eta_0 \\ k_y k_b, & z = 0 \end{cases}$$

where

$$B = \frac{\eta_0}{(p+1)\,\varphi}\left[\frac{1 - k_y k_b + \varphi(p - k_y k_b)}{\sqrt{(p+1)\,\varphi} + \sqrt{(1-k_y k_b)\,(\varphi - 1)}}\right]^2$$

$$y_z(z\cdot H) = k_y k_b{}^{1-z} + \frac{z}{k_v k_t}(1 - k_y k_b)$$

e) $\mu < \frac{p}{p+k_y k_b}, \quad \varphi > \frac{p+1}{p+k_y k_b}$

$$y(z\cdot H, \eta) = \begin{cases} 1/k_b, & z = 1, \quad \eta < \frac{\eta_0(p+k_y k_b)}{\varphi(p+1)}; \\ \sqrt{\frac{(p+1)\,(p+k_y k_b)\,\eta_0}{k_v k_t k_B \eta}} - \frac{p}{k_b}, & z = 1, \quad \frac{\eta_0(p+k_y k_b)}{\varphi(p+1)} < \eta < \frac{\eta_0(p+1)}{\varphi(p+k_y k_b)}; \end{cases}$$

$$y(z \cdot H, \eta) = \begin{cases} k_y, & z = 1, \quad \dfrac{\eta_0(p+1)}{\varphi(p + k_y k_b)} < \eta < \eta_0; \\ k_y k_b, & z = 0; \end{cases}$$

$$y_z(z \cdot H) = k_y k_b^{1-z} + \frac{z}{k_v k_t} (1 - k_y k_b).$$

In cases a), b), c) the bed is completely satiated during adsorption. At the emerging of the flow from the bed during adsorption the concentration of the sorbate leaps from $k_y k_b$ upwards to 1. During desorption there could be present one of the following three possibilities — the concentrations front emerges out of the bed only in part, completely or remains totally inside the layer. In cases d), e) the bed is incompletely satiated during adsorption, the sorbate concentration beyond the bed is constant through the half-cycle and equal to $k_y k_b$. During desorption the front of the concentrations may leave the bed completely or only in part. The working conditions for the bed are determined by ν, k_b and φ. The quantity φ corresponds roughly to the mean velocity of the concentrations front during desorption. If $\varphi < 1$ then in the stationary-cyclic regime the bed is completely satiated during adsorption. At the same time when $\nu < \dfrac{p}{p+1}$, the point in the desorption front where the concentration is maximum has not enough time during the corresponding half-cycle in which to leave the layer. Therefore the concentration of the sorbate in the flow on leaving the bed is constant during the whole half-cycle and is of the maximum value $1/k_b$. If for otherwise equal conditions the length of the bed is smaller than during desorption the concentrations front leaves the bed in part only (case "b") or completely (case "a"). If $\varphi > 1$ and the length of the adsorbent

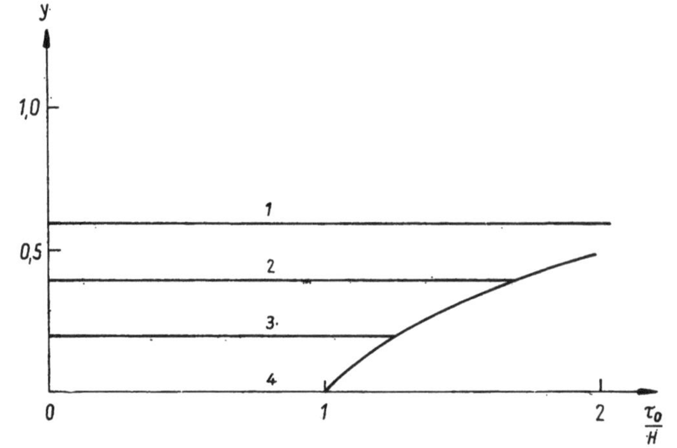

Fig. 2: The mean integral concentration beyond the layer during adsorption in the equilibrium adsorption-desorption process.
$p \to \infty$, $k_v k_t$: I — 0.4; II — 0.6; III — 0.8; IV — 1.

layer is sufficiently large then the satiation of the layer may be incomplete. When this is the case then desorption will be either incomplete (case "d") or complete (case "e").

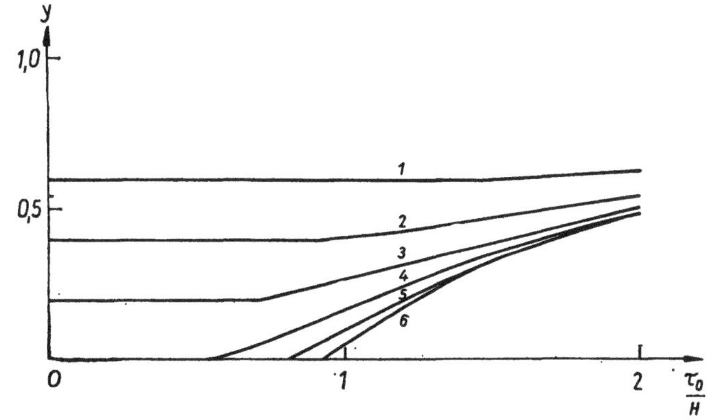

Fig. 3: The mean integral concentration beyond the layer during adsorption in the equilibrium adsorption-desorption process.
$p = 1$, $k_t k_v$: *I* – 0.4; *II* – 0.6; *III* – 0.8; *IV* – 1; *V* – 1.2; *VI* – 1.4

Figs. 2—3 offer the correlations calculated by the obtained formulas depicting in what way the mean integral over every half-cycle concentrations of the sorbate beyond the bed depend on the ratio η_0/H of the flow velocities during adsorption and desorption and on the other parameters of the cyclic process. As it is evident from these pictures the minimum concentration of the sorbate beyond the bed during adsorption is achieved only when $\eta_0/H < 1$, $\varphi \geqq 1$.

With the increase in the curvature of the isotherm the concentration beyond the sorbent layer increases also. The mean concentration beyond the bed during desorption decreases with the increase in η_0/H, flow-velocity and decrease in p.

For numerical calculations of adsorption-desorption processes with arbitrarily taken τ_0 we employed the equations (19—22) for the speedier obtainment of the stationary-cyclical distribution of concentrations. Then the number of cycles until the quasisteady-state regime was reached has not been more than 5 for various sets of parameters. If the isotherm was linear the stationary-cyclical conditions were achieved after 2—3 cycles.

The offered method was used for the study of general regularities of a quasi-steady-state regime. Fig. 4 shows the distributions of concentrations for various values of τ_0. With the increase in τ_0 the functions $y_0(\xi)$ and $y_1(\xi)$ gradually diverge and the curves for $y_0(\xi)$ become more convex. The dotted lines based on the equation (19) represent the in8tial distributions of concentrations in the moving phase, which were used in order to assure the speedier achievement of the quasisteady-state regime. In Fig. 5 there are curves which allow to judge the influence of the isotherm parameter p on the stationary-cyclical distribution of concentrations. It is known that two stages of a frontal sorption

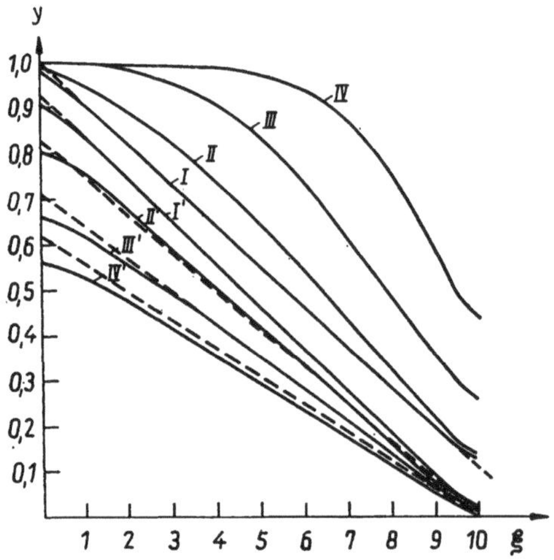

Fig. 4: The influence the half-cycle time on the stationary-cyclical distribution of concentrations in the moving phase.

$p = 0.5$, $\Gamma = 10^4$, $L = 10$, $D = 0.25$, $\varepsilon = 0.4$; $k_\beta = k_v = k_t = 1$, $k_y = 0$.
The end of adsorption: *I, II, III, IV*.
The end of desorption: *I', II', III', IV'*.
The half-cycle time τ_0: *I,I'* — ~ 0; *II, II'* — 1; *III, III'* — 4; *IV, IV'* — 7.
The dotted lines — initial approximations according to (19).

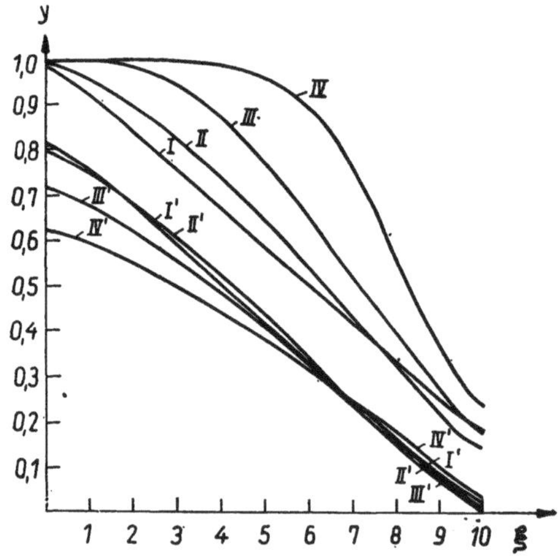

Fig. 5: The influence of the parameter p on the stationary-cyclical distribution of concentrations in a moving phase

$\tau_0 = 1$, $\Gamma = 10^4$, $L = 10$, $D = 0.25$, $\varepsilon = 0.4$, $k_\beta = k_v = k_t = 1$, $k_y = 0$.
The end of adsorption: *I, II, III, IV*.
The end of desorption: *I', II', III', IV'*.
The parameter p: *I, I'* — $\sim \infty$; *II, II'* — 0.5; *III, III'* — 0.1; *IV, IV'* — 0.05.

process could be discerned with $0 < p < \infty$ (convex isotherm): the forming of a quasistationary front and its parallel transfer. A cyclic sorption process has one peculiar feature. If the value of the half-cycle time is not very large the quasistationary front has not time enough to form.

However, the rate of the formation of the front depends considerably on the parameter p, and if p is small the rate of forming is higher. Therefore the curves for the distributions of concentrations at the end of an uneven half-cycle have the largest curvature if p is small. The same curve at the end of an even half-cycle lay lower for small values of p. This could be explained by the fact that when p decreases the desorption of the substance adsorbed by the sorbent

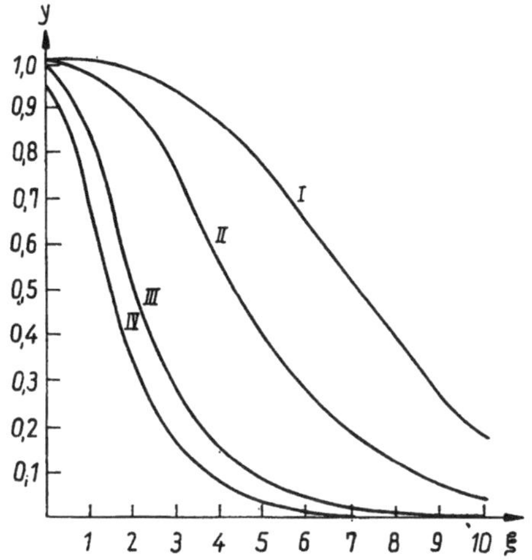

Fig. 6: The distribution of concentrations for the stationary-cyclical regime. The end of adsorption.

$p = 0.1$, $\tau_0 = 1$, $L = 10$, $D = 0.25$, $\varepsilon = 0.4$, $k_v = k_\beta = 1$, $k_y = 0$.
The parameter k_t: *I* – 1; *II* – 1.5; *III* – 3; *IV* – 5.

slows down. Thus if p decreases the curves $y_0(\xi)$ and $y_1(\xi)$ diverge and the concentration of the substance on leaving the bed increases. However, the calculations showed that if $\tau_0/L \ll 1$ there is an optimum value of $p = p_{\text{opt}}$ as regards the exit concentration. The dependence of this value on the other parameters of the process is very faint and this value corresponds to the interval $p > 0.2$, i.e. slightly convex isotherm. The regularity stated above applies to the range of changes $0 < p < p_{\text{opt}}$. If the parameter p increases from p_{opt} to ∞ a gradual slight increase in the exit concentration can be observed. The conclusion can be drawn that the adsorption-desorption process with linear or slightly-curved isotherms is optimum as regards the reduction of the exit concentration. The raising of the parameters k_v, k_t, k_β and the reduction of k_y and k_Γ lead to the decrease of the exit concentration. Fig. 6 shows the stationary-cyclical distributions of concentrations in the moving phase at the end of an uneven half-cycle for various values k_t.

When doing calculations for multicomponent mixtures we employed the equations (19—22) as initial approximations for every component. The number of cycles until the quasisteady-state regime was reached did not exceed in this case ten cycles. Fig. 8 shows the functions of distribution of the concentrations over the length of the layer for a two-component mixture. One can see quite well the effect of cousting a less-adsorbable component.

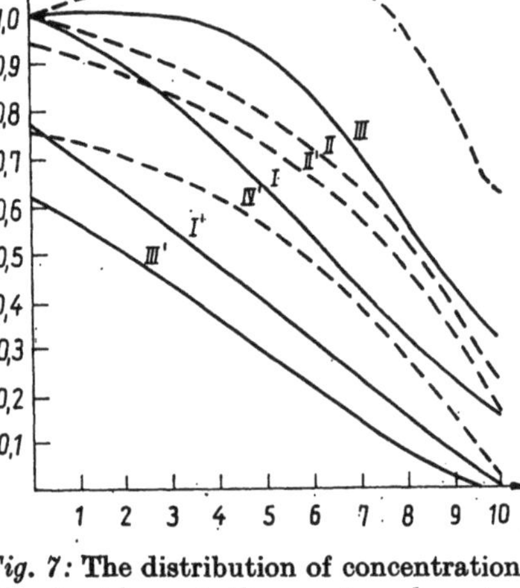

Fig. 7: The distribution of concentrations over the length of the adsorbent layer in the moving phase (unbroken lines) and in the sorbent (dotted lines) for the stationary-cyclical regime when there are two adsorbing substances in the mixture.

$r_2 = 1$; $L = 10$, $D = 0.1$, $\tau_0 = 1$, $p_1 = 0.5$, $p_2 = 2$, $\Gamma = 10^4$, $\delta_2 = 0.1$, $k_{\beta i} = k_v = k_t = 1$; $k_y = 0$ $(i = \overline{1, 2})$.
The end of adsorption: *I, II, III, IV*
The end of desorption: *I', II', III', IV'*
The substance number: *I, II, I', II'* $- i = 1$; *III, IV, III', IV'* $- i = 2$.

The numerical method of calculations introduced in this paper can be employed for engineering calculations and for the optimization of cyclic adsorption-desorption processes. The obtained analytical correlations can be used as initial approximations for the speedier obtainment on IBM of stationary-cyclical distribution of concentrations or as workable formulas for calculation and optimization of short-cycle adsorption-desorption processes.

Designations

$C_i(\xi, \tau)$, $C_i(\zeta, \eta)$ — concentration in the gaseous phase of ith component, kg/m^3
$a_i(\xi, \tau)$, $a_i(\zeta, \eta)$ — concentration in the sorbent of ith component, kg/m^3
$A_{\infty i}$ — limit extent of satiation of the sorbent by i^{th} component, kg/m^3

b_i — constant of LANGMUIR's equation for i^{th} component, m^3/kg
$C_{01}^{(z)}$ — entrance concentration of ith component for the zth half-cycle, kg/m^3
t — time, s
$t^{(z)}$ — time of the zth half-cycle, s
x — coordinate, placed along the adsorber axis, m
$v^{(z)}$ — flow velocity for the zth half-cycle, m/s
$\beta_i^{(z)}$ — mass-transfer coefficient for the zth half-cycle, 1/s
τ_z, η_z — dimensionless time of the zth half-cycle
D^* — effective coefficient of diffusion, m^2/s
ε — porosity

$$\xi = \frac{\beta_1^{(0)}x}{v^{(0)}}, \quad \tau = \frac{\beta_1^{(0)}t}{\Gamma_1}, \quad \zeta = \frac{x\Gamma_1 m}{v^{(0)}t^{(0)}}, \quad \eta = \frac{t}{t^{(0)}}, \quad L = \max_x \xi, \quad H = \max_x \zeta,$$

$$p_i = \frac{1}{b_i C_{0i}^{(0)}}, \quad \Gamma_i = \frac{a_{0i}^{(0)}}{C_{0i}^{(0)}} = \frac{A_{\infty i}}{C_{0i}p_i\left(1 + \sum_{k=1}^{M} \frac{1}{p_k}\right)}, \quad r_i = \frac{\beta_i^{(0)}}{\beta_1^{(0)}}, \quad \delta_i = \frac{\Gamma_i}{\Gamma_1}, \quad m = \frac{1-\varepsilon}{\varepsilon},$$

$$D = \frac{\beta_1^{(0)}D^*}{[v^{(0)}]^2}, \quad y_i = \frac{C_i}{C_{0i}^{(0)}}, \quad \gamma_i = \frac{a_i}{a_{0i}^{(0)}}, \quad k_v = \frac{v^{(1)}}{v^{(0)}}, \quad k_{\beta_i} = \frac{\beta_i^{(1)}}{\beta_i^{(0)}}, \quad k_{y_i} = \frac{C_{0i}^{(1)}}{C_{0i}^{(0)}},$$

$$k_t = \frac{t^{(1)}}{t^{(0)}}, \quad k_\Gamma = \frac{\Gamma^{(1)}}{\Gamma^{(0)}}, \quad k_b = \frac{b^{(1)}}{b^{(0)}}, \quad \alpha = \frac{k_v k_t}{k_\Gamma}, \quad \varphi = \frac{k_v k_t}{k_b}, \quad \varkappa = \frac{k_v}{k_\beta},$$

$$\lambda = \frac{(\alpha - 1)\, m}{\alpha + \varkappa}, \quad \mu = \frac{\eta_0}{H}, \quad \nu = \frac{\varphi\eta_0}{H}.$$

References

[1] KONOBEEV, B. I., FANDEEV, M. A., ARUTYUNYIAN, G. R., NAZAROV, P. S., Theor. Found. Chem. Techn. **10**, № 2, 190 (1976)
[2] BUNKE, G., GELBIN, D., Chem. Enging. Sci. **30**, № 10 (1975)
[3] BEKKER, B. I., TOROCHESHNIKOV, N. S., SHUMYATZKY, Yu. J., Theor. Found. Chem. Techn. **9**, № 5, 771 (1975)
[4] KOLMOGOROV, A. N., FOMIN, C. B., "*Elements of the theory of functions and functional analysis*", Science, Moscow 1968
[5] SAMARSKI, A. A., "*Introduction into the theory of difference schemes*", Science, Moscow 1971

On solving problems of the dynamics of sorption on a digital computer

(A. V. Kolbantsev, V. L. Kolin, I. N. Taganov)

Zusammenfassung:

Gegenstand des Beitrages sind Besonderheiten von Netzmethoden zur Lösung von partiellen Differentialgleichungssystemen für die Dynamik von Sorptionsprozessen. Es wird vorgeschlagen, zur Beschleunigung der Lösung auf EDVA ein ungleichmäßiges Netz mit variablen Grenzen zu verwenden, wobei die Netzknoten im Bereich der stärksten Funktionsänderung am dichtesten sind. Am Beispiel eines nichtlinearen parabolischen Differentialgleichungssystems wird ein Vergleich gleich- und ungleichförmiger Netze durchgeführt und für einen Sonderfall der analytischen Lösung gegenübergestellt. Bei befriedigender Genauigkeit kann die Rechenzeit gegenüber gleichförmigen Netzen auf ein Fünftel gesenkt werden. Hauptanwendungsbereich der vorgeschlagenen Methode sind die Berechnung und Optimierung instationär arbeitender Apparatesysteme.

The problems of the type "the dynamics of sorption" [1] develop as the result of analysis and calculations for various technological processes in the systems "gas – solid matter" and "liquid solid matter". The analytical solutions for the problem of the evolution of the potential field are obtainable only through the introduction of considerable simplifications and mainly for the processes which can be described by linear systems of differential equations in partial derivatives. Undoubtedly, the most effective now is the approximate solution of these problems on a digital computer [2–8].

As a rule, the net method is used when solving numerically the problems of the sorption dynamics type [9–10]. The degree of closeness of an approximate numerical solution to an accurate one depends to a great extent on the choice of the net.

Besides it is possible to employ the evenly spaced nets (with the changes in argument according to the number of the intervals) as well as the unevenly spaced nets. The evenly spaced nets are a bit simpler and require fewer calculations for the values of the functions sought for at one function than the unevenly spaced nets, and so are more frequently used. However, in a number of cases the evenness of the net is a considerable drawback, which calls for a great increase in the calculation time for all the area for which the values of function are set.

Indeed, in order to ensure the required accuracy in various areas according to the form of the function various sizes of the steps in the net may be needed.

When employing evenly-spaced nets one is obliged to choose the size of the steps so that the required accuracy is ensured for the area of the maximum changes in the defined function. But thus some rather lengthy parts of the area for which the values of function are set are calculated with unjustifiably small step and high degree of accuracy and the overall time of the solution of the problem becomes rather long.

The use of unevenly spaced nets (with the steps varying along the coordinate) can eliminate the above-mentioned defect. But the calculations of the finite differences equations, which approximate differential equations, is more complex in the unevenly constructed nets. Moreover, the use of such nets becomes complicated because the position of the areas where the function is defined and for which the varying steps can be employed is changed during the calculation process.

No matter whether the concentration front is quasi-stationary or contracts or becomes washed-out with the movement along the sorbent bed the boundaries of the area with the higher gradient of concentration have in various moments of time different co-ordinates [6].

It is obvious that when solving such problems with evenly spaced nets a considerable part of the area where the function is defined is calculated by the same small steps as are necessary for a much smaller area, which is characterized by a bigger gradient of concentration. This leads to considerable losses in the computer time.

However, now we have already great experience in solving the problems of the dynamics of sorption, and further in employing the solutions for explorations of the general laws of the development of the processes and for engineering purposes. It should be noted, however, that the computer time required prevents us from the consideration of such problems as real cyclic sorption-desorption processes leading to the stationary-cyclic condition, calculations for the work of multiapparatus chemico-technological systems, which include nonstationary working apparatus, such as adsorbers, desorbers, ion-exchange columns. We are also still prevented from finding computer solutions to such problems of optimization as finding the constants of the equations included in mathematical models by the available experimental data, the discrimination between hypothetical factors which determine the process velocity; we cannot yet find the parameters of a process either which will ensure its best technological or economic indices.

In order to further develop and widen the area of useful mathematical methods in the solving, calculating and improvement of non-stationary processes of the "dynamics of sorption" type, speedier methods of numerical solution for the systems of differential equation by a digital computer have to be developed.

Some acceleration can be obtained if the calculation of a non-stationary field of concentrations in the process of sorption is reduced to the problem of the definition of a concentrations front with mobile boundaries of separate areas as in the processes which are accompanied by phase changes [11].

Then for each area of concentrations separated within the front the evenly-spaced net can be employed.

The auxiliary conditions for the inner boundaries of the areas should be taken as to assure a continuous and smooth function, that is:

$$\varphi|_{b+0} = \mathrm{q}|_{b-0} \qquad \nabla\varphi|_{b+0} = \nabla\varphi|_{b-0}.$$

This refinement of the whole problem by dividing the area where the function is defined into several sub-areas makes it possible to choose for every sub-area an evenly (or unevenly) spaced net with a step which will ensure the required accuracy, and the overall number of the net functions as well as the computer-time are reduced as compared to the methods usually employed.

Let us take a closer look at the above-described method of approach e.g. by the following mathematical model of an isothermal sorption process:

$$D\varepsilon \frac{\partial^2 c}{\partial x^2} - V \frac{\partial c}{\partial x} - (1-\varepsilon) \frac{\partial a}{\partial t} \varepsilon = \frac{\partial c}{\partial t} \tag{1}$$

$$\frac{\partial a}{\partial t} = f(c, a). \tag{2}$$

Let us assume that the initial and boundary conditions satisfy the following correlations:

$$\begin{aligned} &t = 0 \quad c(x) = 0, \qquad a(y) = a_{\text{init}} \\ &x = 0 \quad VC = VC|_{x=0^+} - D \frac{\partial c}{\partial x}\bigg|_{x=0^+} \\ &x = L \quad \frac{\partial c}{\partial x}\bigg|_{x=L^-} = 0. \end{aligned} \tag{3}$$

Let us introduce the following non-dimensional parameters and constants:

$$D = \frac{D \cdot \beta}{V^2} \varepsilon^2; \quad G = \frac{a_0}{c_{in}}; \quad \alpha = \frac{1-\varepsilon}{\varepsilon};$$

$$y = \frac{c}{c_{in}}; \quad \gamma = \frac{a}{a_0}; \quad \xi = \frac{\beta x}{V} \varepsilon; \quad \tau = \frac{\beta \cdot t}{G}.$$

Then instead of (1)—(3) we have

$$D \frac{\partial^2 y}{\partial \xi^2} - \frac{\partial y}{\partial \xi} - \alpha \frac{\partial \gamma}{\partial \tau} = \frac{1}{G} \frac{\partial y}{\partial \tau} \tag{4}$$

$$\frac{\partial \gamma}{\partial \tau} = f_1(y, \gamma) \tag{5}$$

$$\begin{aligned} &\tau = 0 \quad \gamma(\xi) = 0, \quad \gamma(\xi) = 0 \\ &\xi = 0 \quad y_{in} = y|_{\xi=0^+} - D \frac{\partial y}{\partial \xi}\bigg|_{\xi=0^+} \\ &\xi = L_1 \quad \frac{\partial y}{\partial \xi}\bigg|_{\xi=L_1^-} = 0. \end{aligned} \tag{6}$$

As the law of change for $y(\xi, \tau)$ and $\gamma(\xi, \tau)$ is as shown in Fig. 1, we divide the area of the definition for the function in three parts in the following way:

$$\xi \in [0, X_1] \qquad |y(\xi) - y(0)| \leqq E_{11} \qquad |\gamma(\xi) - \gamma(0)| \leqq E_{12} \tag{7}$$

$$\xi \in [X_2, L_1] \qquad |y(\xi) - y(L_1)| \leqq E_{21} \qquad |\gamma(\xi) - \gamma(L_1)| \leqq E_{22} \tag{8}$$

$$\xi \in [X_1, X_2] \qquad E_{21} + y(L)_1 < y(\xi) < y(0) - E_{11} \qquad E_{22} + \gamma(L_1) < \gamma(\xi) < \gamma(0) - E_{21}. \tag{9}$$

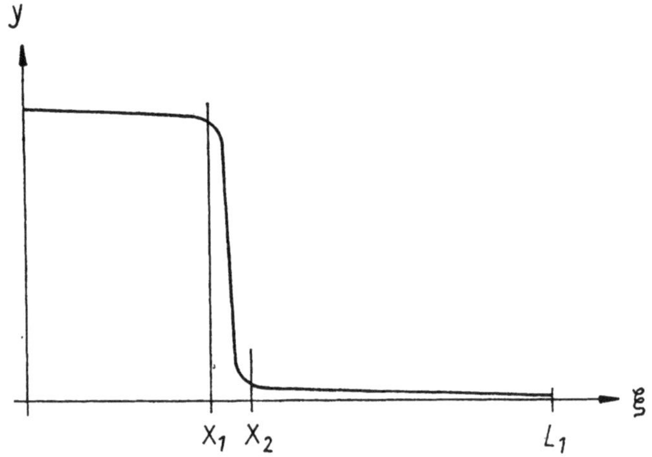

Fig. 1: Concentration profile.

Thus we obtain the following system of equations and boundary conditions:

$$\left.\begin{aligned} \frac{1}{G}\frac{\partial y}{\partial \tau} &= D\frac{\partial^2 y}{\partial \xi} - \frac{\partial y}{\partial \xi} - \alpha\frac{\partial \gamma}{\partial \tau} \\ \frac{\partial y}{\partial \tau} &= f_1(y, \gamma) \end{aligned}\right\} \text{ when } \begin{aligned} &\xi \in (0, X_1) \\ &\xi \in (X_1, X_2) \\ &\xi \in (X_2, L_1) \end{aligned} \tag{10}$$

$$y_{in} = y|_{\xi=0^+} - D\left.\frac{\partial y}{\partial \xi}\right|_{\xi=0_+} \qquad \text{when } \xi = 0 \tag{11}$$

$$y|_{X_1-0} = y|_{X_1+0}; \quad \left.\frac{\partial y}{\partial \xi}\right|_{X_1-0} = \left.\frac{\partial y}{\partial \xi}\right|_{X_1+0} \qquad \text{when } \xi = X_1 \tag{12}$$

$$y|_{X_2-0} = \gamma|_{X_2+0}; \quad \left.\frac{\partial y}{\partial \xi}\right|_{X_2-0} = \left.\frac{\partial y}{\partial \xi}\right|_{X_2-0} \qquad \text{when } \xi = X_2 \tag{13}$$

$$\left.\frac{\partial y}{\partial \xi}\right|_{L_1^-} = 0 \qquad \text{when } \xi = L_1 \tag{14}$$

and the above-written conditions (7), (8) for the definition of the position of the boundaries $X_1(\tau)$, $X_2(\tau)$.

Now we will introduce the new independent variables θ and η, which are connected with ξ, τ by the following correlations:

$$\theta = \tau, \quad \eta = \frac{\xi - X_{l.b.}}{X_{r.b.} - X_{l.b.}}. \tag{15}$$

After going over to new variables the system (10) and boundary conditions (10)—(16) will look as follows:

$$\begin{aligned} &\frac{D}{(X_{r.b.} - X_{l.b.})^2} \frac{\partial^2 y}{\partial \eta^2} - \frac{1}{X_{r.b.} - X_{l.b.}} \frac{\partial y}{\partial \eta} - \alpha \frac{\partial y}{\partial \theta} \\ &+ \alpha \left[\eta \frac{dX_{r.b.}}{d\theta} + (1 - \eta) \frac{dX_{l.b.}}{d\theta}\right] \frac{1}{X_{r.b.} - X_{l.b.}} \frac{\partial \gamma}{\partial \eta} = \frac{1}{G} \frac{\partial y}{\partial \theta} \\ &- \frac{1}{G} \left[\eta \frac{dX_{r.b.}}{d\theta} + (1 - \eta) \frac{dX_{l.b.}}{d\theta}\right] \frac{1}{X_{r.b.} - X_{l.b.}} \cdot \frac{\partial y}{\partial \eta} \end{aligned} \tag{16}$$

$$\frac{\partial \gamma}{\partial \theta} - \frac{\eta \dfrac{dX_{r.b.}}{d\theta} + (1 - \eta) \dfrac{dX_{l.b.}}{d\theta}}{X_{r.b.} - X_{l.b.}} \frac{\partial \gamma}{\partial \eta} = f_1(y, \gamma)$$

$$\begin{cases} \eta = 0 \quad y_{in} = \gamma|_{\eta=0} - D \dfrac{1}{X_1} \dfrac{\partial y}{\partial \eta}\Big|_{\eta=0} \\ \eta = 1 \quad y|_{X_1-0} = y|_{X_1+0}; \ \dfrac{1}{X_1} \dfrac{\partial y}{\partial \eta}\Big|_{X_1-0} = \dfrac{1}{X_2 - X_1} \dfrac{\partial y}{\partial \eta}\Big|_{X_1+0} \end{cases} \quad \text{1st area} \tag{17}$$

$$\begin{cases} \eta = 0 \quad y|_{X_1-0} = y|_{X_1+0}; \ \dfrac{1}{X_1} \dfrac{\partial y}{\partial \eta}\Big|_{X_1-0} = \dfrac{1}{X_2 - X_1} \dfrac{\partial y}{\partial \eta}\Big|_{X_1+0} \\ \eta = 1 \quad y|_{X_2-0} = y|_{X_2+0}; \ \dfrac{1}{X_2 - X_1} \dfrac{\partial y}{\partial \eta}\Big|_{X_2-0} = \dfrac{1}{L_1 - X_2} \dfrac{\partial y}{\partial \eta}\Big|_{X_2+0} \end{cases} \quad \text{2nd area} \tag{18}$$

$$\begin{cases} \eta = 0 \quad y|_{X_2-0} = y|_{X_2+0}; \ \dfrac{1}{X_2 - X_1} \dfrac{\partial y}{\partial \eta}\Big|_{X_2-0} = \dfrac{1}{L_1 - X_2} \dfrac{\partial y}{\partial \eta}\Big|_{X_2+0} \\ \eta = 1 \quad \dfrac{\partial y}{\partial \eta}\Big|_{L^-} = 0. \end{cases} \quad \text{3rd area} \tag{19}$$

Thus for each area (Fig. 1) in the η, θ co-ordinates we can choose an evenly-spaced net which will ensure the required accuracy of the numerical solution. The virtue of such approach lies in a fixed position of the net functions in the η, θ coordinates.

In this work we used for the solution of the system (16) — (19) a three-points implicit scheme (Fig. 2). After the replacement of the derivatives by finite differences, we get for each of the sub-areas, shown in Fig. 1, the following

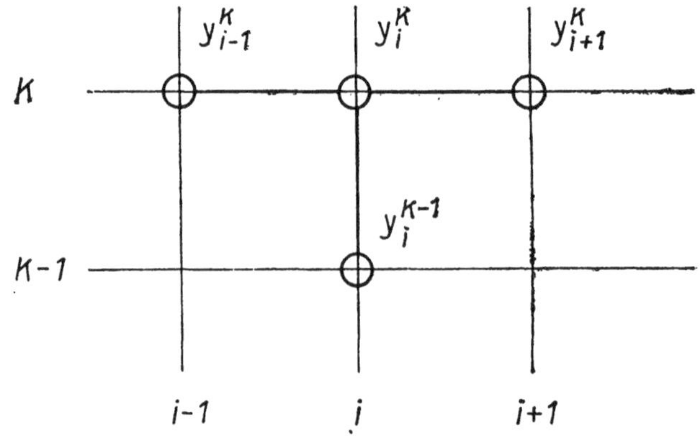

Fig. 2: Scheme of finite difference approximation.

system of algebraic equations

$$A_i^k y_{i+1}^k - B_i^k y_i^k - C_i^k y_{i-1}^k = -F_i^{k-1} \tag{20}$$

$$\gamma_i^k = \gamma_i^{k-1} + \Delta\theta \left[f_1(y_1^{k-1}, \gamma_i^{k-1}) + \frac{i\Delta\eta \dfrac{dX^{k-1}_{r.b.}}{d\theta} + (1 - i\Delta\eta) \dfrac{dX^{k-1}_{l.b.}}{d\theta}}{X^{k-1}_{r.b.} - X^{k-1}_{l.b.}} \times \frac{\gamma^{k-1}_{i+1} - \gamma_i^{k-1}}{\Delta\eta} \right], \quad i = 1, 2, \ldots, n-1. \tag{21}$$

$$i = n \quad \gamma_i^k = \gamma_n^{k-1} + \Delta\theta \cdot f_1(Y_n^{k-1}, \gamma_n^{k-1}) \tag{22}$$

where

$$A_i^k = \frac{D}{(X^{k-1}_{r.b.} - X^{k-1}_{l.b.})^2} \cdot \frac{1}{\Delta\eta^2} - \frac{1}{X^{k-1}_{r.b.} - X^{k-1}_{l.b.}} \cdot \frac{1}{2\Delta\eta} + \frac{1}{G} \cdot \frac{1}{2\Delta\eta} \cdot \frac{1}{X^{k-1}_{r.b.} - X^{k-1}_{l.b.}} \left[\Delta\eta \cdot i \cdot \frac{dX^{k-1}_{r.b.}}{d\theta} + (1 - \Delta\eta \cdot i) \frac{dX^{x-1}_{l.b.}}{d\theta} \right]$$

$$B_i^k = \frac{1}{G\Delta\theta} + \frac{1}{\Delta\eta^2} \cdot \frac{2D}{(X^{k-1}_{r.b.} - X^{k-1}_{l.b.})^2}$$

$$C_i^k = \frac{1}{\Delta\eta^2} \cdot \frac{D}{(X^{k-1}_{r.b.} - X^{k-1}_{l.b.})^2} + \frac{1}{2\Delta\eta} \cdot \frac{1}{X_{r.b.} - X_{l.b.}} - \frac{1}{2G\Delta\eta} \cdot \frac{1}{X^{k-1}_{r.b.} - X^{k-1}_{l.b.}} \left[\Delta\eta \cdot i \cdot \frac{dX^{k-1}_{r.b.}}{d\theta} + (1 - \Delta\eta \cdot i) \frac{dX^{k-1}}{d\theta} \right]$$

$$F_i^k = \frac{1}{G\Delta\theta} y_i^{k-1} - f_1(y_i^{k-1}, \gamma_i^{k-1}).$$

$\Delta\theta = \frac{\theta_{in}}{l}$ – step in the θ coordinate;

l – the number of junctions in the net;

n – the number of junctions in the η coordinate;

$\Delta\eta = \frac{1}{n}$.

We solved the system (20)–(22) completed by two equations for boundary conditions by a back-run method [12].

For the example which will be considered further we assumed that at the initial moment of time the sorbent layer is completely free of the sorbate and its entrance concentration in the flow C_{in} = const. is constant during the whole process. The kinetic function $f(c, a)$ we took as

$$f(c, a) = \beta C \left(1 - \frac{a}{a_0}\right) \tag{23}$$

what is we assumed that the isotherm of sorption is sharply convex (Γ – shaped).

It should be noted that for such an approach to the problem of the dynamics of sorption the form of the function $f_1(y, \gamma)$ does not matter and the results discussed there are of rather a general character. Changes in initial and boundary conditions also do not introduce any considerable differences into the following algorithm.

Then the non-dimensional form of (5) is as follows:

$$\frac{\partial\gamma}{\partial\tau} = f_1(y, \gamma) = y(1 - \gamma). \tag{24}$$

The coefficients are defined by the equations

$$P_{i+1} = \frac{A_k^1}{B_i^k - C_i^k P_i}, \quad Q_{i+1} = \frac{F_i^k + C_i^k Q_i}{B_i^k - C_i^k P_i} \tag{25}$$

$$(i = 1, 2, \ldots, n - 1).$$

The values of y we get from the equation

$$y_{i-1}^k = P_i y_i^k + Q_i \qquad (i = n, n - 1, \ldots, 1) \tag{26}$$

and the values of γ_i^k from (21). The values of P_1, Q_1, y_n^k are determined by the boundary conditions:

1st area:

$$P_1 = \frac{\frac{D}{X_1^k \Delta\eta}}{1 + \frac{D}{X_1^k \cdot \Delta\eta}}; \quad Q_1 = \frac{1}{1 + \frac{D}{X_1^k \Delta\eta}}; \quad y_n^k = y^k|_{X_1^k} \tag{27}$$

2nd area:

$$P_1 = \frac{1}{1 + \frac{X_2^k - X_1^k}{X_1^k} \cdot \frac{\Delta\eta}{\Delta\eta|_{1az}} (1 - P_n|_{1az})} \tag{28}$$

$$Q_1 = -\frac{\frac{X_2^k - X_1^k}{X_1^k} \cdot \frac{\Delta\eta}{\Delta\eta|_{1az}} \cdot Q_n|_{1az}}{1 + \frac{X_2^k - X_1^k}{X_1^k} \frac{\Delta\eta}{\Delta\eta|_{1az}} (1 - P_n|_{1az})} \tag{29}$$

$$y_n^k = y^k|_{X_2^k} \tag{30}$$

3rd area:

$$P_1 = \frac{1}{1 + \frac{L - X_2^k}{X_2^k - X_1^k} \frac{\Delta\eta}{\Delta\eta|_{2az}} (1 - P_n|_{2az})} \tag{31}$$

$$Q_1 = -\frac{\frac{L - X_2^k}{X_2^k - X_1^k} \cdot \frac{\Delta\eta}{\Delta\eta|_{2az}} \cdot Q_n|_{2az}}{1 + \frac{L - X_2^k}{X_2^k - X_1^k} \cdot \frac{\Delta\eta}{\Delta\eta|_{2az}} (1 - P_n|_{2az})} \tag{32}$$

$$y_n^k = \frac{Q_n}{1 - P_n}. \tag{33}$$

When compiling the algorithm for this problem for the digital computer we took into account that the rates of movement for the left and right boundaries (Fig. 1), which are included in A_i, C_i, were previously unknown to us. So we chose the initial numerical values for these rates so that the change of the position of these boundaries in the θ, η co-ordinates would be a value less than small ε. For further approximate evaluations of $\frac{dX_{r.b.}}{d\theta}$ and $\frac{dX_{l.b.}}{d\theta}$ we used the values for the previous step by θ.

In order to evaluate the efficiency of the offered approximate method the problem has been solved with the use of the uneven net. The implicit 3-point scheme has also been employed (Fig. 2). After substituting the finite differences instead of the derivatives we obtained

$$Ay_{i+1}^k - By_i^k + Cy_{i-1}^k = -F \tag{34}$$

$$\gamma_i^k = \gamma_i^{k-1} + \Delta\tau \cdot f_1(y_i^{k-1}, \gamma_i^{k-1}) \tag{35}$$

$$A = \frac{D}{\Delta\xi^2} - \frac{1}{2\Delta\xi}; \quad B = \frac{2D}{\Delta\xi^2} + \frac{1}{G} \frac{1}{\Delta\tau};$$

$$C = \frac{D}{\Delta\xi^2} + \frac{1}{2\Delta\xi}; \quad F = \frac{1}{G} \cdot \frac{1}{\Delta\tau} \cdot y_i^{k-1} - f_1(y_i^{k-1}, \gamma_i^{k-1}).$$

The coefficients were defined by the formula

$$P_{i+1} = \frac{A}{B - CP_i}; \quad Q_{i+1} = \frac{F + CQ_i}{B - CP_i} \quad (i = 1, 2, \ldots, n - 1). \quad (36)$$

The back-run was carried out by the formula

$$y_{i-1}^k = P_i r_i^k + Q_i. \quad (37)$$

The values of P_1, Q_1, y_n^k are determined by the boundary conditions

$$P_1 = \frac{\frac{D}{\Delta\xi}}{1 + \frac{D}{\Delta\xi}}; \quad Q_1 = \frac{1}{1 + \frac{D}{\Delta\xi}} \quad (38)$$

$$y_n^k = \frac{Q_n}{1 - P_n}. \quad (39)$$

Here:

$\Delta\xi = \frac{L}{n}$

n — number of points order the coordinate ξ;
$\Delta\tau$ — step in time.

In order to obtain comparable results of the calculations, before printing $y(\xi, \tau)$, $\gamma(\xi, \tau)$ the interpolation of the net function was carried out by the Aitken scheme (a polynomial of the 3rd order was used). Printed were the values of the coordinates ξ, τ which correspond to the set numerical values of relative concentrations $y(\xi, \tau)$, $\gamma(\xi, \tau)$.

In table 1 there is the example of the comparison between the results at the initial stage of the process, when the concentrations front is unformed yet and the use of the unevenly-spaced net with a small number of functions in the area I (Fig. 1) leads to the largest errors. As this comparison shows the method of calculation offered can ensure the same degree of accuracy as an evenly spaced net, while the number of functions required is 5 times less than with an evenly spaced net.

As the numerical experiments showed, the employment of the offered method, while the accuracy of the obtained results is the same, takes about 5 times less time than when the evenly-spaced net is used. A non-linear system of differential equations with partial derivatives (4), (24) with zero initial data and boundary conditions corresponding to a frontal sorption process ($y_{in} = 1$; $(\gamma(\xi, 0) = y(\xi, 0) = 0)$, when $D \to 0$, can be solved analytically and such solutions were obtained in [6]. This makes possible a comparison between the accurate and approximate numerical solutions and thus evaluates once more the accuracy of the offered speedier method of solution of the problem of the "dynamics of sorption" type on a digital computer. The analytical solution of

Table 1: Comparison of the results of the numerical solution of the problem when employing evenly and unevenly spaced nets in mobile co-ordinates

y	ξ calculated			
	evenly-spaced net		unevenly-spaced net	
	number of junctions			
	210	150	10 + 30 + 10	5 + 15 + 10
$\tau = 0.1$				
0.9	0	0	0	0
0.7	0	0	0	0
0.5	0.3130	0.3154	0.3142	0.3154
0.3	0.8360	0.8394	0.8376	0.8394
0.1	1.948	1.953	1.947	1.950
0.01	4.259	4.270	4.259	4.265
0.0001	8.874	8.893	8.762	8.829
$\tau = 0.5$				
0.9	0	0	0	0
0.7	0.01849	0.02015	0.01922	0.02014
0.5	0.4166	0.4191	0.4175	0.4191
0.3	0.9856	0.9895	0.9870	0.9894
0.1	2.140	2.146	2.140	2.143
0.01	4.473	4.481	4.474	4.479
0.0001	9.086	9.107	8.992	9.009
$\tau = 1.0$				
0.9	0	0	0	0
0.7	0.1139	0.1131	0.1134	0.1127
0.5	0.5791	0.5782	0.5788	0.5762
0.3	1.203	1.203	1.203	1.200
0.1	2.404	2.405	2.404	2.400
0.01	4.753	4.759	4.757	4,752
0.0001	9.371	9.386	9.284	9.270

the problem ($D \to 0$) for the $y(\xi, \tau)$ distribution is a follows:

$$y(\xi, \tau) = \frac{1}{1 + e^{-\tau}(e^{\xi} - 1)} \tag{40}$$

where, the non-dimensional variables now are:

$$\xi = \frac{\beta x}{V} \cdot \varepsilon \qquad \tau = \frac{\beta\left(t - \dfrac{x}{V}\right)}{G},$$

and the kinetic coefficient β introduced earlier is related to a unit of volume of the sorbent layer.

If we assume $x/V \to 0$, $f_{sp.} = 1$, $\varepsilon = 0.5$ and $D \to 0$, the system of differential equations (4), (24) with the above-stated initial and boundary conditions yields the solution (40).

The results of the calculations by the method offered and analytical solution (40) are compared in the table.

The differences between the values of relative concentrations y obtained by the approximate numerical method and by analytical solution are rather small (not more than 0.6% rel.).

Table 2: Comparison of the results obtained by numerical and analytical methods

$G = 10^3$; $D = 5 \cdot 10^{-8}$; $\Delta\tau = 0.005$; $n = 50$ (10 + 30 + 10).

Time	Coordinate	Concentration, y	
τ		Solution	
		Analytical	Numerical
0.1	0.05680	0.9458	0.95
	0.1159	0.8999	0.9
	0.3874	0.7002	0.7
	0.7443	0.5001	0.5
	1.276	0.2997	0.3
	2.399	0.09943	0.1
	9.354	$0.9573 \cdot 10^{-4}$	$1 \cdot 10^{-4}$
0.2	0.06228	0.95001	0.95
	0.1269	0.90025	0.9
	0.4196	0.70083	0.7
	0.7965	0.50072	0.5
	1.347	0.30028	0.3
	2.487	0.09971	0.1
	9.451	$0.9601 \cdot 10^{-4}$	$1 \cdot 10^{-4}$
0.5	0.09077	0.95044	0.95
	0.1827	0.90089	0.9
	0.5727	0.70213	0.7
	1.031	0.5025	0.5
	1.651	0.3019	0.3
	2.852	0.1004	0.1
	9.843	$0.9662 \cdot 10^{-4}$	$1 \cdot 10^{-4}$
1.0	0.1324	0.9505	0.95
	0.2614	0.9020	0.9
	0.7659	0.7025	0.7
	1.304	0.5032	0.5
	1.983	0.3027	0.3
	3.228	0.1008	0.1
	10.24	$0.9729 \cdot 10^{-4}$	$1 \cdot 10^{-4}$

Taking into account the approximate nature of the mathematical models of the sorption processes, the accuracy of the evaluation of the constants defining the equilibrium of the kinetics of sorption, the hydrodynamical structure of the flow in the bed and requirements as to the accuracy of the engineering calculations, the results should be considered satisfactory.

Thus the method offered allows to obtain solutions of the problems of sorption dynamics with sufficient accuracy and also ensures considerable economy in computer time.

References

[1] RACHINSKI, V. V., "*Introduction into General Theory of Adsorption Dynamics and Chromatography*", Nauka, Moscow, 1962
[2] ACRIVOS, A., Ind. Eng. Chem. **48**, 703 (1956)
[3] CARTER, I. B., Chem. and Process Eng. **47**, 70 (1966)
[4] CARTER, I. B., Trans. Instn. Chem. Engrs. **44**, 253 (1966)
[5] CHI, C. W., WASAU, D. T., AIChE Journal **16**, 23 (1970)
[6] KOLIN, V. L., BAGROV, G. I., MUSAKIN, G. A., SEMENOV, V. P., "*Kinetics and Dynamics of Physical Adsorption*", Nauka, Moscow 1973
[7] MEYER, O. A., WEBER, T. B., AIChE Journal **13**, 457 (1967)
[8] FROLOV, V. F., LESIN, Ju. S., "*Kinetics and Dynamics of Physical Adsorption*", Nauka, Moscow 1973
[9] SAMARSKI, A. A., "*Introduction into the Theory of Difference Schemes*", Science, Moscow 1971
[10] GODUNOV, S. K., RYABENKI, V. S., "*Difference Schemes — Introduction into the Theory*", Science, Moscow 1971
[11] RUBINSTEIN, I. L., "*About the Solution of the* STEFAN-*Problem*", Sinatne, Riga 1967
[12] DEMIDOVITCH, B. P., MAROM, I. A., SHUWALOVA, E. S., "*Numerical Methods of Analysis*", Nauka, Moscow 1967

Mathematical simulation of sorption processes in solid moving beds

(A. A. SEBALLO, V. L. KOLIN, I. N. TAGANOV, N. S. FROLOVA)

Zusammenfassung:

Es werden mathematische Modelle für die isotherme und nichtisotherme Gegen- und Kreuzstromadsorption einzelner Komponenten in der Wanderschicht formuliert und Gleichungen für die Geschwindigkeit der Sorptionswelle abgeleitet. Für Prozesse mit linearen Isothermen werden numerische Berechnungsmethoden für konstante und veränderliche Stofftransportkoeffizienten aufgestellt. Auf Grund der Rechenergebnisse wird der Einfluß kinetischer Eigenschaften untersucht und der Einfluß der freiwerdenden Wärme auf die Form der Komponentenverteilung in der Phasen und Gasphase berücksichtigt. Die grundlegenden Beziehungen, die bei der Berechnung der Adsorption einer Komponente verwendet werden, wurden auf die Berechnung der Adsorption von Gemischen und die Verdrängungsdesorption ausgedehnt. Die Rechnerergebnisse werden mit experimentellen Daten verglichen.

1. Introduction

The development of chemical and oil refining industries is accompanied by the improvement of hydrocarbon separation techniques and the creation of continuous processes. Solid moving beds of adsorbents are particularly attractive for these purposes because they are characterized by economical metal consumption, continuous operation, low specific consumption of adsorbents per key component unit and make possible total process mechanization and automatic control.

Continuous processes of hydrocarbon separation in moving adsorbent beds are widely used in different branches of industry [1—9]. However, studies of adsorption dynamics in moving beds and methods of its calculation are rather scarce [10—13]. Besides, solutions were chiefly obtained for the simplest individual cases without considering the effects of process nonisothermality, axial dispersion and multiplicity of adsorbed components.

A great variety of adsorbents and adsorptives in different separation and purification processes prompted the necessity of studying general dynamic laws for single and multicomponent adsorption in moving beds and methods of their calculation.

2. Isothermal dynamic pure component adsorption from the flow of inert carrier gas

Consider that a column into which a gas flow with a constant initial concentration of adsorptive C_0 is fed at $x = 0$ at a linear velocity u the exhausted adsorbent leaving the column at a linear velocity w. If the adsorption heat effects, nonuniform motion of the grains and longitudinal diffusion in the carrier flow are neglected, the material balance equations for each phase in the first approximation have the following form:

$$\frac{\partial c}{\partial t} = -u \frac{\partial c}{\partial x} - f(c, a)$$
$$\frac{\partial a}{\partial t} = w \frac{\partial a}{\partial x} + f(c, a). \tag{1}$$

The boundary and initial conditions for the adsorption process of the fresh adsorbent are:

$$t = 0 \quad x > 0 \qquad c = 0 \quad a = 0 \tag{2}$$
$$x = 0 \quad t \geqq 0 \qquad c = c_0 \quad a = a_0 \tag{3}$$
$$x = \infty \quad 0 \leqq t < \infty \qquad c = 0 \quad a = 0 \tag{4}$$

Problem (1)—(4) was formulated by TODES [10]. It presented a non-linear system of equations and the analytical solution of the problem has not been obtained yet. Therefore we have confined ourselves to the analysis of the asymptotic stage ($t \to \infty$). Introducing a new variable $z = x - vt$ and transforming the partial derivatives in (1) give

$$\frac{\partial c}{\partial x} = \frac{dc}{dz}; \quad \frac{\partial z}{\partial x} = \frac{dc}{dz}; \quad \frac{\partial a}{\partial x} = \frac{da}{dz} \tag{5}$$

$$\frac{\partial c}{\partial t} = \frac{\partial c}{\partial z}; \quad \frac{\partial z}{\partial t} = -v \frac{\partial c}{\partial z}; \quad \frac{\partial a}{\partial t} = -v \frac{da}{dz}. \tag{6}$$

Substitution of (6) for (1) gives

$$-v \frac{dc}{dz} = v \frac{dc}{dz} - f(c, a)$$
$$-v \frac{da}{dz} = w \frac{da}{dz} + f(c, a) \tag{7}$$

With equation (5) the boundary conditions become

$$z = -\infty \qquad c = c_0 \qquad a = a_0 \tag{8}$$
$$z = +\infty \qquad c = 0 \qquad a = 0 \tag{9}$$

Simple transformations give the expression of stationary front motion:

$$v = \frac{uc_0 - wa_0}{c_0 + a_0} = (u + w) \frac{c_0}{c_0 + a_0} - w. \tag{10}$$

If the longitudinal dispersion is considered, system (1) becomes:

$$\frac{\partial c}{\partial t} = -u \frac{\partial c}{\partial x} - f(c, a) + D^* \frac{\partial^2 c}{\partial x^2}$$
$$\frac{\partial a}{\partial t} = w \frac{\partial a}{\partial x} + f(c, a). \tag{11}$$

In this case the boundary and initial conditions may be written as follows:

$$t = 0 \quad x = 0 \quad c = 0 \quad a = 0 \tag{12}$$

$$x = 0 \quad t \geqq 0 \quad c = c_0 \quad a = a_0 \quad \frac{\partial c}{\partial x} = 0 \tag{13}$$

$$x = \infty \quad 0 \leqq t \leqq \infty \quad c = 0 \quad a = 0 \quad \frac{\partial c}{\partial x} = 0. \tag{14}$$

On introducing the variable z and after simple rearrangement we obtain

$$v = \frac{uc_0 - wa_0}{a_0 + c_0} = (u + w) \frac{c_0}{a_0 + c_0} - w. \tag{15}$$

Thus the stationary front velocity at $D^* \neq 0$ is defined by the same formula as that for front velocity at $D^* = 0$. From (15) the maximal solid phase velocity is defined, at which the sorption wave is carried from the apparatus ($v < 0$), i.e. parallel transfer of the concentration front becomes impossible with

$$w_{\max} = \frac{c_0}{a_0} u \frac{c_0 + a_0}{a_0} v_0 = \frac{v_0}{1 - v_0/u} \tag{16}$$

here $v_0 = u \dfrac{c_0}{a_0 + c_0}$ is the stationary concentration front velocity in the fixed bed. Since $a_0 \gg c_0$ and $v_0 \ll u$ ... only slightly exceeds v_0. It is known that at the sufficiently high velocity of the adsorbent counterflow ($w \gg w_{\max}$) all existing methods of analysis for unsteady-state adsorption processes are useless. Since apparatuses with moving beds are designed for a continuous mode of operation and the nonstationary stage is of limited importance we shall analyse only steady-state processes.

Now we consider a steady-state adsorption process in a column with the bed length H. Fresh adsorbent ($a = 0$) is fed from the top of the column at the velocity w. Gas (liquid) enters the column at $x = 0$ at the velocity u. The key component concentration is c_0. The adsorption kinetics is defined by the expression $\dfrac{da}{dt} = \beta[c - \varphi(a)]$. If one neglects the longitudinal dispersion and non-uniform motion of the adsorbent grains, which is true at $D/d > 10$, and assumes isothermality of the process in the first approximation, the mass balance equations and the boundary conditions for each phase participating in mass transfer have the following form:

$$u \frac{dc}{dx} = -\beta[c - \varphi(a)]$$
$$w \frac{da}{dx} = -\beta[c - \varphi(a)] \tag{17}$$

$$x = 0 \qquad c = c_0 \tag{18}$$

$$x = H \qquad a = a_0. \tag{19}$$

The general solution of problems (17)—(19) for any adsorption isotherm and adsorption kinetics was obtained by TODES [10]

$$\int_{c_0}^{c_k} \frac{dc}{f\left[c_1 \frac{u}{w}(c - c_k)\right]} = -\frac{x}{u}. \tag{20}$$

Here the integration constant c_k is defined from

$$\int_{c_0}^{c} \frac{dc}{f\left[c_1 \frac{u}{w}(c - c_k)\right]} = \frac{H}{u}. \tag{21}$$

c_k can be determined from (21) when the concrete forms of the kinetic and isotherm equations are known. In the case of the non-linear isotherm the integration of (21) is sometimes possible (for instance; for the LANGMUIR isotherm). This, however, would yield the definition of c_k in the form of a cumbersome transcendental equation.

Now we consider the situation for the case of the linear isotherm. After introducing the dimensionless parameters and variables

$$y = \frac{c}{c_0}; \quad Y = \frac{a}{a_0}; \quad \xi = \frac{x\beta}{u}; \quad \psi = \frac{a_0 w}{c_0 u} \tag{22}$$
$$L = \frac{H\beta}{u}; \quad \varphi(Y) = \frac{\varphi(a)}{c_0}$$

we rewrite (17)—(19) in the form

$$\frac{dy}{d\xi} = [\varphi(Y) - y] \tag{23}$$
$$\frac{dY}{d\xi} = \frac{1}{\psi}[\varphi(Y) - y]$$

$$\xi = 0 \qquad y = 1 \tag{24}$$

$$\xi = L \qquad Y = 0. \tag{25}$$

Then solution (23) takes the form

$$y = \frac{e^{\left(\frac{1}{\psi}-1\right)L}}{\psi - e^{\left(\frac{1}{\psi}-1\right)L}} \left[\psi e^{\left(\frac{1}{\psi}-1\right)(\xi-L)} - 1\right] \tag{26}$$

$$Y = \frac{e^{\left(\frac{1}{\psi}-1\right)L}}{\psi - e^{\left(\frac{1}{\psi}-1\right)L}} \left[e^{\left(\frac{1}{\psi}-1\right)(\xi-L)} - 1\right]. \tag{27}$$

In the case of the non-linear isotherm and conditions (24)—(25) system (23) present a two-point boundary problem, which cannot be solved by direct integration methods. At first it should be reduced to the initial value problem. Multiplying the second equation of system (23) by ψ and substracting this from the first equation give

$$\frac{dy}{d\xi} = \psi \frac{dY}{d\xi}. \tag{28}$$

Integration of (28) leads to

$$Y_H = \frac{1}{\psi}(y - k) \tag{29}$$

here k — the integration constant equal to y_k.

From condition (24)

$$Y_H = \frac{1}{\psi}(1 - y_k) \tag{30}$$

where Y_H is the saturated fraction of the adsorbent leaving the column. Since $y_k \ll 1$ it is neglected in the first approximation, and then

$$Y(0) = Y_H = \begin{cases} 1/\psi & \psi > 1 \\ 1 & \psi \leqq 1. \end{cases} \tag{31}$$

Thus we get the lacking boundary condition for the determination of the distribution functions of $y(\xi)$ and $Y(\xi)$ from system (23) with the initial conditions

$$\xi = 0 \quad y = 1 \quad Y = \begin{cases} 1/\psi & \psi > 1 \\ 1 & \psi \leqq 1. \end{cases} \tag{32}$$

Now we shall analyse the adsorption problem, in which the longitudinal diffusion effects are considered. The mass balance equations have the form

$$u\frac{dc}{dx} = -\beta[c - \varphi(a)] + D^* \frac{d^2c}{dx^2}$$

$$-w\frac{da}{dx} = -\beta[c - \mathrm{q}(a)]. \tag{33}$$

Fresh adsorbent ($a = 0$) is fed from the top of the column ($x = H$) and some quantity of adsorptive uc_0 is fed from the bottom of the column ($x = 0$). The adsorptive is transferred through the column by gas-carrier flow $uc(0)$ and longitudinal diffusion $-D^* \frac{dc(0)}{dx}$ then the boundary conditions have the form:

$$x = 0 \qquad uc_0 = uc(0) - D^* \frac{dc(0)}{dx} \tag{34}$$

$$x = H \quad D^* = \frac{dc(H)}{dx} = 0; \quad a = 0. \tag{35}$$

On introducing dimensionless parameters and variables (33)—(35) may be rewritten

$$\frac{dy}{d\xi} = z$$

$$\frac{dz}{d\xi} = G[z + y + \varphi(Y)] \tag{36}$$

$$\frac{dY}{d\xi} = \frac{1}{\psi}[\varphi(Y) - y]$$

$$\xi = 0 \qquad y = 1 + \frac{1}{G} z \tag{37}$$

$$\xi = L \qquad Y = 0; \quad \frac{1}{G} z = 0. \tag{38}$$

For the HENRY type isotherm the analytical solution of (36)—(38) reads:

$$y = \varkappa \left[1 + \frac{1 + G\psi K_2}{G(K_1 - K_2)} e^{\left(K_1 G + \frac{1}{\psi}\right)(\xi - L)} - \frac{1 + G\psi K_1}{G(K_1 - K_2)} e^{\left(K_2 G + \frac{1}{\psi}\right)(\xi - L)}\right] \tag{39}$$

$$Y = \varkappa \left[1 + \frac{1 + GK_2}{G(K_1 - K_2)} e^{\left(K_1 G + \frac{1}{\psi}\right)(\xi - L)} - \frac{1 + GK_1}{G(K_1 - K_2)} e^{\left(K_2 G + \frac{1}{\psi}\right)(\xi - L)}\right] \tag{40}$$

where

$$x = \frac{e^{\left(K_2 G + \frac{1}{\psi}\right)L}}{e^{\left(K_2 G + \frac{1}{\psi}\right)L} + \psi \frac{1 + GK_2}{G(K_1 - K_2)} e^{-(K_1 - K_2)GL} - \psi \frac{1 + GK_1}{G(K_1 - K_2)}} \tag{41}$$

$$K_1 = \frac{1}{2}\left(1 - \frac{1}{G\psi}\right) + \sqrt{\frac{1}{\psi}\left(1 - \frac{1}{G\psi}\right)^2 + \frac{1}{G}}$$

$$K_2 = \frac{1}{2}\left(1 - \frac{1}{G\psi}\right) - \sqrt{\frac{1}{\psi}\left(1 - \frac{1}{G\psi}\right)^2 + \frac{1}{G}}.$$

For the case of adsorption in the adsorbent bed with some residual adsorptive loading the solution of system (36)—(38) has the form

$$
\begin{aligned}
y &= c_1 + c_2 e^{\lambda_2 \xi} + c_3 e^{\lambda \xi} \\
z &= c_2 \lambda e^{\lambda_2 \xi} + c_3 \lambda_3 e^{\lambda_3 \xi} \\
Y &= c_1 + c_2 \frac{e\lambda_2 \xi}{1 - \lambda_2 \xi} + c_3 \frac{e\lambda_3 \xi}{1 - \lambda_3 \psi} \\
\lambda_2 &= \frac{1}{2}\left(G + \frac{1}{\psi}\right) + \sqrt{\frac{1}{\psi}\left(G - \frac{1}{\psi}\right) + G} \\
\lambda_3 &= \frac{1}{2}\left(G + \frac{1}{\psi}\right) - \sqrt{\frac{1}{\psi}\left(G - \frac{1}{\psi}\right) + G} \\
c_1 &= 1 + c_3 \left[e^{(\lambda_3 - \lambda_2)L} - 1 - \frac{\lambda_3}{G} e^{(\lambda_3 - \lambda_2)L} + \frac{\lambda_3}{G}\right] \\
c_2 &= -c_3 \frac{\lambda_3}{\lambda_2} e^{(\lambda_3 - \lambda_2)L} \\
c_3 &= \frac{Y_{0CT} - 1}{\frac{\lambda_3}{\lambda_2} e^{(\lambda_3 - \lambda_2)} - 1 - \frac{\lambda_3}{G} e^{(\lambda_3 - \lambda_2)L} + \frac{\lambda_3}{G} - \frac{\lambda_3 e^{\lambda_3 L}}{\lambda_2 (1 - \lambda_3 \psi)} + \frac{e^{\lambda_2 L}}{1 - \lambda_2 L}}.
\end{aligned}
\tag{42}
$$

By letting $D^* \to 0$ (or $G \to \infty$) and removing the uncertainty $K_2 G = \frac{K_2}{1/G} = \frac{0}{0}$ with the L'HOSPITAL rule we get the expressions

$$
y = \frac{e^{\left(\frac{1}{\psi} - 1\right)L}}{\psi - e^{\left(\frac{1}{\psi} - 1\right)L}} \left[\psi e^{\left(\frac{1}{\psi} - 1\right)(\xi - L)} - 1\right] \tag{43}
$$

$$
Y = \frac{e^{\left(\frac{1}{\psi} - 1\right)L}}{\psi - e^{\left(\frac{1}{\psi} - 1\right)L}} \left[e^{\left(\frac{1}{\psi} - 1\right)(\xi - L)} - 1\right] \tag{44}
$$

similar to the equations derived by direct integration of system (23) with condition (24)—(25). Analysis of expressions (4), (44) shows exponential drop of the component concentration in gas and solid phase along the bed length (Fig. 1).

As it is mentioned above system (36)—(38) presents a two-point boundary problem and the attempts to solve this analytically are still unsuccessful. Numerical methods for calculation of such problems are not properly developed either. However, the numerical solution of system (36) can be obtained if we take into account that according to the physical sense

$$
Y > 0; \quad Y < 1; \quad y < 1 \tag{45}
$$

the numerical solution can be obtained by the method of successive approximations [14]. Since at $\xi = L$ two conditions are given it is reasonable to integrate between L and 0 with a fixed value of the relative breakthrough concentration at the bed exit. As a first approximation for y_n' we may use the solution for the linear adsorption isotherm — the value of $y(L)$ from equation (39). When $y_n^{(i)}$ is known, system (36) can be calculated by integration between L

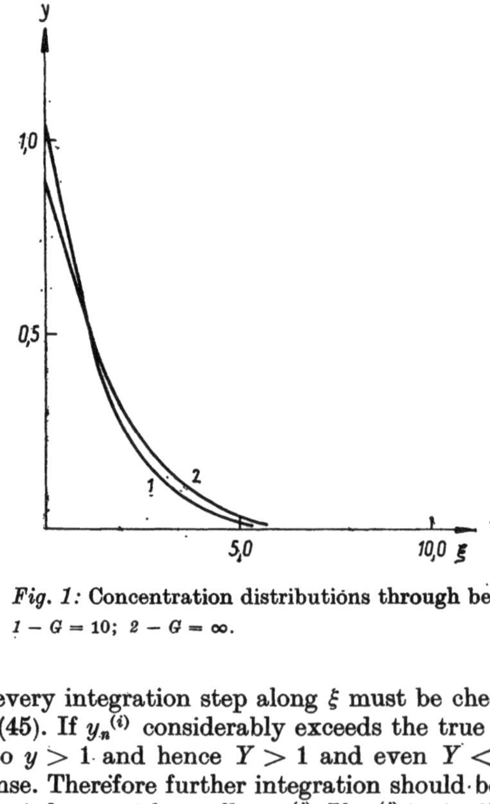

Fig. 1: Concentration distributions through bed length.
1 — $G = 10$; 2 — $G = \infty$.

and 0 but every integration step along ξ must be checked for satisfiability of conditions (45). If $y_n^{(i)}$ considerably exceeds the true value y_r the calculation may lead to $y > 1$ and hence $Y > 1$ and even $Y < 0$ which is deprived of physical sense. Therefore further integration should be interrupted and calculation be carried out with smaller $y_n^{(i)}$. If $y_n^{(i)}$ is significantly smaller than the true value, one should go over to calculation with a greater value $y_n^{(i)}$. In case conditions (45) are satisfied and integration proceeds to the end comparison is to be made of $y(0)$ found in integration with $y^T(0)$ derived from boundary condition (37), in the following expression:

$$\Delta = y(0) - y^T(0) = y(0) - 1 - \frac{1}{G} z(0). \tag{46}$$

The possibility of ceasing iterations is tested with the formula

$$w = \left| \frac{y_n^{(i)} - y_n^{(i-1)}}{y_n^{(i)}} \right| \leq \varepsilon.$$

With $\Delta > 0$, $w > \varepsilon$ the calculation accuracy is of the order of $10^{-2} \div 10^{-3}$; the calculation should be carried out with a smaller $y_n^{(i)}$. When $\Delta < 0$ and $w > \varepsilon$ large values $y_n^{(i)}$ are used. At $w \leqq \varepsilon$ the calculation is considered to be completed. The method converges in the same way as that of cutting the segment in half.

From above it follows that the mass-transfer coefficient is constant. In case that the relation $\beta = f(a/a_0) = f(Y)$ is valid the dimensionless variable

$$g(Y) = \frac{\beta(Y)}{\beta_{cp}} \tag{47}$$

$$\beta_{cp} = \frac{1}{Y_H - Y_k} \int_{Y_k}^{Y_H} \beta(Y)\, dY$$

is introduced. The calculation method remains the same. In programming these algorithms the adsorption isotherms and the right-hand parts of differential equations are evaluated as separate subroutines. This mode provides the possibility of solving problems for any adsorption isotherm by the substitution of corresponding procedures. In the calculation of thea bove mentioned problems it is often convenient to combine different techniques for the sake of machine time saving. One such combination is described in [15].

When it is necessary to determine the bed length, in which the adsorptive concentration drops from an initial to some breakthrough value (c_{np}) a new system of defining $x = L - \xi$ is introduced. In this case the system of differential equations and boundary conditions take the form:

$$\frac{dz_1}{dx} = z_2 \tag{48}$$

$$\frac{dz_2}{dx} = G[z_1 - \varphi(z_3)] \tag{49}$$

$$\frac{dz_3}{dx} = \frac{1}{\psi}[z_1 - \varphi(z_3)]$$

$$x = L \qquad z_1 = 1 - \frac{1}{G} z_2 \tag{50}$$

$$x = 0 \qquad \frac{1}{G} z_2 = 0; \quad z_3 = 0. \tag{51}$$

Substitution of (51) by

$$x = 0; \quad -\frac{1}{G} z_2 = 0; \quad z_3 = 0; \quad z_1 = \frac{0.5 c_{np}}{c_0} \tag{52}$$

and integration of system (49) under condition (52) by Runge-Kutta method until

$$z_1 - 1 + \frac{1}{G} z_2 = 0 \tag{53}$$

yield the apparatus length providing a fixed breakthrough concentration. At the calculation accuracy of 1% and integration step $L/100$ the integration process converges in 2—3 iterations.

The results of digital computer simulation show (Fig. 2) a large effect of axial dispersion on the distribution in the range of low relative concentrations, which should be considered in designing processes of fine purification. Calculation accuracy increases when one considers the dependence of the mass-transfer coefficient on the extent of adsorption space filling (introduction of the varible mass-transfer coefficient in the equations).

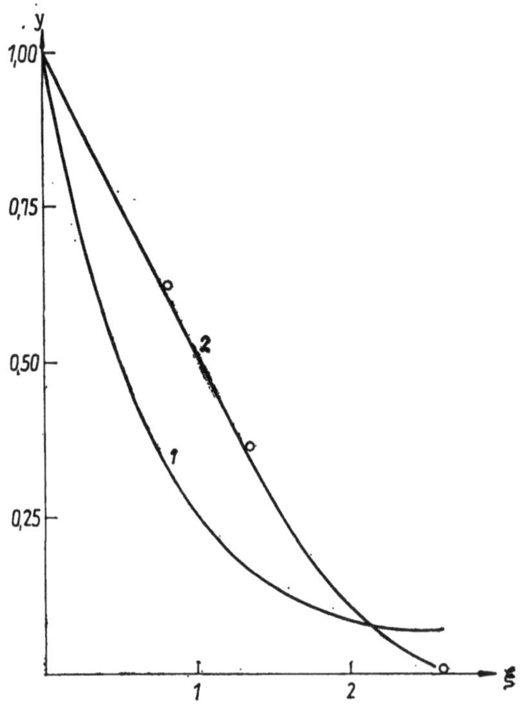

Fig. 2: Concentration distributions through bed length. Experimental data are indicated by dots.

$1 - \beta = \text{const.}; \quad 2 - \beta = f\left(\frac{a}{a_0}\right).$

3. Non-isothermal dynamic pure component adsorption from inert carrier flow

Adsorption is known to be an exothermic process. The evolution of adsorption heat causes temperature rise in adsorbent and gas (liquid) flow and affects the stationary distribution of concentrations. When adsorption takes place from diluted flows or in columns of small diameter, evolving heat is lost to the environment and the process may be regarded approximately isothermal. If concentrations of adsorptive are fairly high, the bed heats considerably and the adsorption dynamics in this case is characterized by the presence of heat and adsorption waves. The temperature rise in the adsorption zone is accompanied by the dropping of the adsorptive relative vapour pressure and equilibrium adsorption value. The overall effect of these factors greatly reduces the driving force.

Now we regard the adsorption process on a bed of uniform cross-section with phase counterflow. In deriving the mass balance equations for the nonisothermal adsorption process, the same simplifying assumptions are made as for the isothermal problem. In deriving the heat balance equations it is assumed that:

1. bed particles are so small or their heat conductance so high that in any spot of the particle the temperature is the same;
2. heat transfer along the bed is performed by gas flow and axial heat conductance of the solid phase may be neglected;
3. the thermophysical characteristics of the system remain constant in the process;
4. the environmental temperature and the initial gas flow temperature are arbitrarily assumed to be equal to 0;
5. the gas and adsorbent temperature at a given bed section is constant, i.e. adsorption heat is spent on heating of the gas and column walls and as environmental losses. In this case the material balance equation for the bed particles per particle volume unit has the form:

$$h_s(1 - f)\frac{\partial T_s}{\partial t} = wh_s\frac{\partial T_s}{\partial x} + q\beta[\dot{c} - \varphi(a, T_s)] - k(T_s - T_g) \qquad (54)$$

where:

h_s — the volume heat capacity of the bed;
w — the adsorbent flow velocity, cm/s;
f — the void fraction of the bed;
T_g, T_s — the gas and bed temperatures at a given moment of time at the distance x from the gas entrance in the adsorption column;
q — the adsorption heat, ccal/g;
K — the volume coefficient of heat transfer from the gas flow to the bed particles.

Thus the heat quantity change in the bed per time unit is equal to the sum of the heat quantity brought by the solid phase flow and that transferred from the grains to the gas flow. The heat balance equation for the gas in a bed volume unit is:

$$h_g f\frac{\partial T_g}{\partial t} = K(T_s - T_g) - uh_g\frac{\partial T_g}{\partial x} - \frac{2\alpha_s}{r}[T_g - T_c], \qquad (55)$$

where:

h_g — the specific volume heat capacity of the gas;
α_s — the heat transfer coefficient from the gas to the column wall;
r — the adsorption column radius, cm;
T_c — the temperature of the column wall at a given moment and at the distance x from the gas entrance.

The heat balance equation for the adsorption column wall per surface area unit has the following form:

$$\delta h_c \frac{\partial T_c}{\partial t} = \alpha_s(T_q - T_s) - \alpha_c T_c \tag{56}$$

here h_c — the volume heat capacity of the wall material; α_c — the heat transfer coefficient from the walls to the environment; δ — the wall thickness. Hence, using the approximations assumed in [16] we derive the fundamental equations defining the gas and bed temperature changes with time at a given bed section

$$h_s(1 - f) \frac{\partial T_s}{\partial t} = wh_s \frac{\partial T_s}{\partial x} + q\beta[c - \varphi(a, T_s)] - K(T_s - T_g) \tag{57}$$

$$h_g \frac{\partial T_q}{\partial t} = K(T_s - T_g) - uh_g \frac{\partial T_q}{\partial x} - \frac{2\alpha_n}{r} T_g \tag{58}$$

where

$$h = h_g f + \frac{2\delta}{r} h_c \frac{\alpha_s}{\alpha_s + \alpha_c}; \quad \alpha_n = \frac{\alpha_s \cdot \alpha_c}{\alpha_s + \alpha_c}.$$

Thus equations (17), (18) and (57), (58) and the initial and boundary conditions

$$t = 0 \qquad c = a = T_g = T_s = 0 \tag{59}$$

$$t \geqq 0 \begin{cases} x = 0 & c = c_0 & T_q = 0 \\ x = H & a = 0 & T_s = 0 \end{cases} \tag{60}$$

give the description of adsorption in a moving bed, in which the evolving heat effect is considered. For a steday-state process $\frac{\partial T_s}{\partial t} = \frac{\partial T_g}{\partial t} = 0$. Finally on introducing the dimensionless parameters and variables

$$\xi = \frac{x\beta}{u}; \quad y = \frac{c}{c_0}; \quad Y = \frac{a}{a_0}; \quad \theta_g \doteq \frac{T_g}{T_0}; \quad \theta_s \doteq \frac{T_s}{T_0};$$

$$\frac{\varphi(a, T_s)}{c_0} = \varphi(Y, \theta_s); \; L = \frac{H\beta}{u}; \quad N = \frac{KU}{wh_s \cdot \beta};$$

$$\Omega = \frac{c_0 u_q}{w h_s T_0}; \quad D = \frac{c_0 \cdot u}{a_0 \cdot w}; \quad F = \frac{K}{h_q \cdot \beta}; \quad \Phi = \frac{2\alpha_n}{r \cdot h_q \cdot \beta}$$

we get the system of differential equations and boundary conditions

$$\begin{aligned} \frac{dy}{d\xi} &= \varphi(Y, \theta_s) - y \\ \frac{dY}{d\xi} &= D[\varphi(Y, \theta_s) - y] \end{aligned} \tag{61}$$

$$\frac{d\theta_s}{d\xi} = N(\theta_s - \theta_g) + [\varphi(Y, \theta_s) - y]$$
$$\frac{d\theta_q}{d\xi} = F(\theta_s - \theta_q) - \Phi\theta_q \tag{62}$$

$$\xi = 0 \quad y = 1; \quad \theta_q = 0$$
$$\xi = L \quad Y = 0; \quad \theta_s = 0. \tag{63}$$

The adsorption isotherm is defined by the DUBININ-RADUSHKEVITCH equation, i.e. it is assumed that

$$\varphi(Y, \theta) = \frac{\varphi(a, T)}{c_0} = \frac{p}{1 + \theta} \exp\left[\frac{B^*\theta - (qD_1 \ln Y)^{1/2}}{1 + \theta}\right]$$

where

$$p_0 = \frac{c_s}{c_0}; \quad q = \left(\ln \frac{p}{p_0}\right)^2; \quad B^* = \frac{\lambda}{RT_0}; \quad D_1 = \frac{\beta \cdot 2.3^2}{T_0^2 \cdot B}. \tag{64}$$

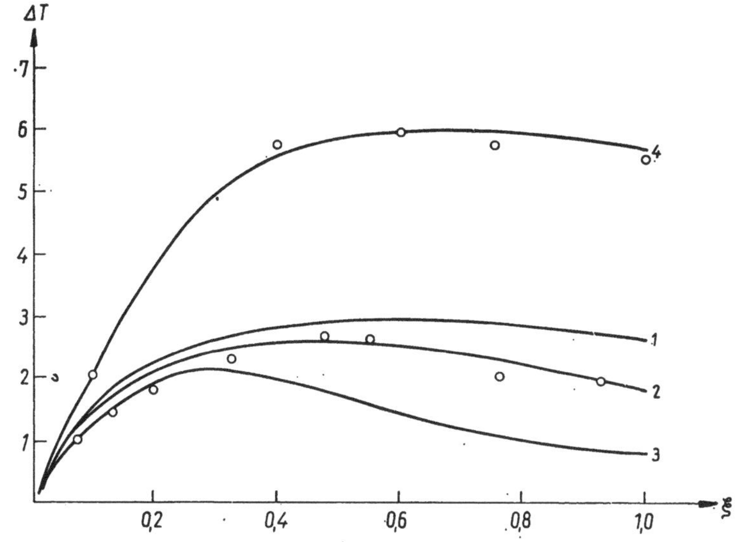

Fig. 3: Temperature distributions through the bed length $w = 0.324 \cdot 10^{-2}$ cm/s, benzene concentration 5.3 g/m³.
1 — $\alpha_n = 2.9$; 2 — $\alpha_n = 8.5$;
3 — $\alpha_n = 21$. Benzene concentration 9.7 g/m³.
4 — $\alpha_n = 21$. Experimental data are indicated by the dots.

The technique in [18] is used for the description of thermophysical characteristics.

Mathematical simulation on a digital computer shows that the increase of the heat transfer coefficient, k, more than twice has no essential effect on the form of the temperature distribution curves but results merely in their slight shift. A somewhat different character of change is observed for the heat transfer coefficient α_n (Fig. 3). The rise of the heat transfer coefficient is followed by distortion of the heat wave in the middle and terminal beds of the adsorption column. With small α_n only the frontal part of the bed is heated. In that case owing to minor heat losses some constant temperature characteristic for given operating conditions is installed practically through the whole column. With

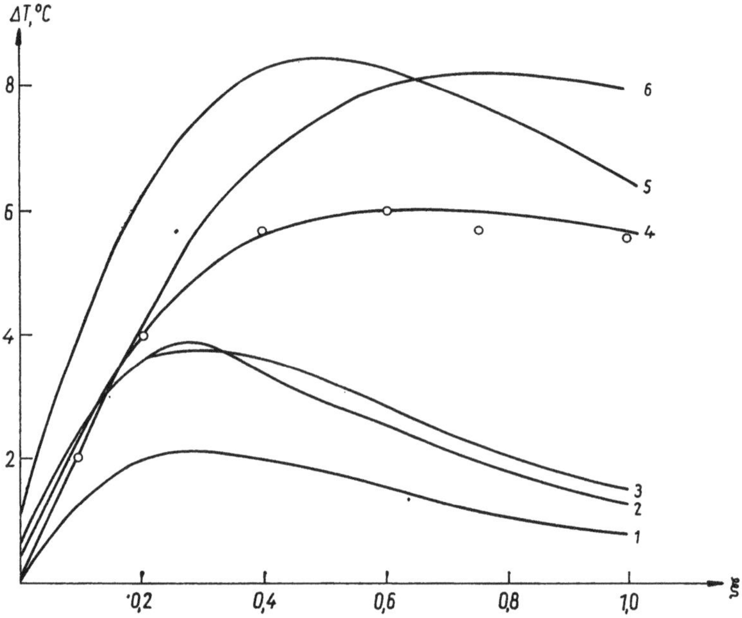

Fig. 4: Temperature distributions through the bed length. Designations are the same as in fig. 3.

$u = 28.1$ cm/s; $k = 2085$.
1 — $c = 5.3$ g/m³; *2* — $c = 9.7$ g/m³, $w = 0.705 \cdot 10^{-2}$; $K = 838$.
4 — $c = 9.7$, $u = 28.1$, $w = 0.324 \cdot 10^{-2}$.
6 — $c = 16.0$, $u = 28.1$. *5* — $c = 16.0$, $u = 590$.

the intense heating this results in the adsorption capacity decrease and lengthening of the adsorption zone. When the heat transfer through the column walls is rapid (curve 2) we observe temperature rise in the front of the column and subsequent temperature drop in the middle and terminal beds. In that case the middle and closing beds are working under nearly isothermic conditions. An even sharper temperature drop is observed when $\alpha_n = 0.5 \cdot 10^{-6}$ (curve 3). In practice one is usually faced with some intermediate cases (curve 2) or processes similar to curve 4. According to the calculation results the growth of the concentration from 5.3 to 9.7 g/m³ is followed by the bed temperature rise and shift of the distribution curve along the bed (Fig. 4, curve 13). A similar relation is observed when the flow rate of the adsorbent increases (curves

2.3). With the growth of the adsorbent flow rate the temperature distribution curve is shifted to the left and that results in the temperature drop in the middle and final parts of the adsorption column. When the gas flow rate increases together with the adsorptive quantity fed to the bed per time unit, the heat transfer through the column becomes more intensive. As a result the heat front shifts to the right and practically the whole column (except the frontal areas of the bed) works under non-isothermal conditions.

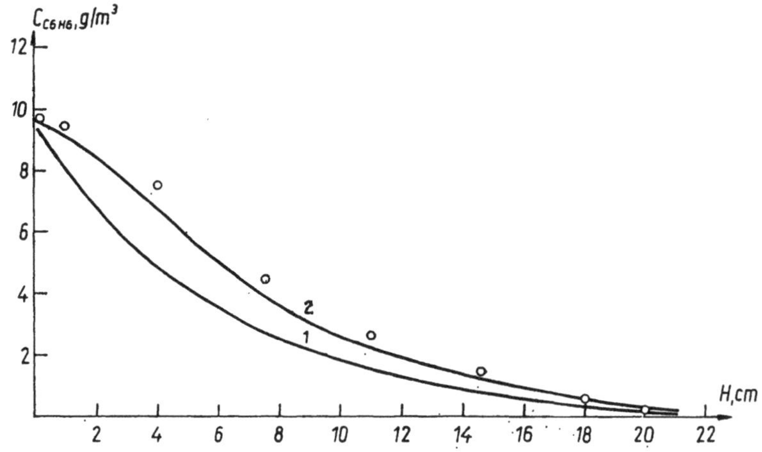

Fig. 5: Concentration distributions through bed length.

u – isothermic model

2 – non-isothermic model

1 = 28.1 cm/s.; c_0 = 9.7 g/m³; w = 0.324 · 10^{-2} cm/s. α_n = 37.7; K = 838.

The comparison of the data in Fig. 5 shows that the concentration distribution curves calculated according to the isothermal model lie much lower than those obtained according to the non-isothermal model. The greatest discrepancy is observed in the higher and average concentration range when the adsorption isotherm is less steep and the evolving heat effect most obvious. The results of calculation according to the non-siothermal model are in good agreement with the experiment. This agreement forms a basis for computer simulation of operating conditions and apparatus dimensions, which provide the required efficiency of purification of gas-vapour flow. The calculated temperature distribution through the bed length make it possible to determine the zone of the most intensive heat evolution and to choose the conditions for the better heat transfer from that zone.

4. Isothermal dynamic adsorption from mixtures

The theoretical approach to the definition of the dynamic mixed adsorption was similar to that described in [17]. In a steady-state process concentrations of components do not change in time. Fresh adsorbent is fed at the linear flow rate w from the top of an adsorption column at $x = H$, where H is the column height. Mixture of n components with the concentrations c_{i0} is fed from the bottom of the column at $x = 0$. The adsorption heat effect being neglected, one can assume that the adsorption rate is directly proportional to the difference between the local component concentration in the carrier flow and the concentration $c_{ip} = \varphi(a_1, a_2, \ldots, a_n)$ in equilibrium with the amount adsorbed a_i.

Hence the mass balance equations for component i may be written in the following form:

$$u \frac{dc_i}{dx} = -\beta_i[c_i - \varphi_i(a_1, a_2, \ldots, a_n) + D^* \frac{d^2c_i}{dx^2}$$
$$-w \frac{da_i}{dx} = \beta_i[c_i - \varphi_i(a_1, a_2, \ldots, a_n)]. \tag{65}$$

The boundary conditions for component c_i have the form

$$x = 0 \qquad uc_{i0} = uc_i(0) - D^* \frac{dc_i(0)}{dx} \tag{66}$$

$$x = H \qquad a_i = 0 \qquad D^* \frac{dc_i(H)}{dx} = 0. \tag{67}$$

On introducing the dimensionless parameters and variables:

$$y_i = \frac{c_i}{c_{0i}}; \quad Y_i = \frac{a_i}{a_{0i}}; \quad G = \frac{u^2}{\beta_1 D^*}; \quad \psi_i = \frac{a_{i0} \cdot w}{c_{i0} \cdot u}$$
$$\varphi_i(Y_1, Y_2, \ldots, Y_n) = \frac{\varphi_i(a_1, a_2, \ldots, a_n)}{c_{i0}}; \quad \xi = \frac{x\beta_1}{u};$$
$$L = \frac{H\beta}{u}; \quad \Phi_i = \frac{\beta_1}{\beta_i}; \quad \frac{dy_i}{d\xi} = z_i;$$

the system of the differential equations and the boundary conditions can be rewritten in the following form:

$$\frac{dy_i}{d\xi} = z_i$$
$$\frac{dz_i}{d\xi} = G\left\{z_i + \frac{1}{\Phi_i}[y_i - \varphi_i(Y_1, Y_2, \ldots, Y_n)]\right\} \tag{68}$$
$$\frac{dY_i}{d\xi} = \frac{1}{\Phi_i \psi_i}[\varphi_i(Y_1, Y_2, \ldots, Y_n)]$$

$$\xi = 0 \qquad y_i = 1 + \frac{1}{G} z_i \tag{69}$$

$$\xi = L \qquad Y_i = 0; \qquad \frac{1}{G} z = 0. \tag{70}$$

In case of the linear adsorption isotherm system (68) transforms into:

$$\begin{aligned} \frac{dy_i}{d\xi} &= z_i \\ \frac{dz_i}{d\xi} &= G\left\{z_i + \frac{1}{\Phi_i}[y_i - Y_i]\right\} \\ \frac{dY_i}{d\xi} &= \frac{1}{\Phi_i\psi_i}[Y_i - y_i]. \end{aligned} \tag{71}$$

This is a linear system describing adsorption of i independently adsorbed components. The system can be solved in analytical form. On multiplying the second equation by $\left(-\frac{1}{G}\right)$ and the third by $(-\varphi_i)$ and adding them up one gets:

$$\frac{d}{d\xi}\left[y_i - \frac{1}{G}z_i - \varphi_i Y_i\right] = 0. \tag{72}$$

Multiplying the second equation by K the third by m and adding them up one gets:

$$\frac{d}{d\xi}[y_i + kz_i + mY_i] = \left(\frac{KG}{\Phi_i} - \frac{m}{\Phi_i\psi_i}\right)y_i + (1 + KG)\,z_i + \left(\frac{m}{\Phi_i\varphi_i} - \frac{KG}{\Phi_i}\right)Y_i. \tag{73}$$

Transformation of the right-hand part of equation (73) into the linear function of $y_i + kz_i + mY_i$ gives

$$\frac{d}{d\xi}[y_i + kz_i + mY_i] = \left(\frac{KG}{\Phi_i} - \frac{m}{\psi_i\Phi_i}\right)\left[y_i + \frac{1 + KG}{\frac{KG}{\Phi_i} - \frac{m}{\Phi_i\psi_i}} - Y_i\right]. \tag{74}$$

From (74) follows that assuming

$$\frac{1 + KG}{\frac{KG}{\Phi_i} - \frac{m}{\Phi_i\psi_i}} = k \qquad m = -1 \tag{75}$$

one gets

$$\frac{d}{d\xi}[y_i + kz_i - Y_i] = ce^{-\frac{1}{\Phi_i}\left(KG + \frac{1}{\psi_i}\right)\xi}. \tag{76}$$

Integration of (76) gives

$$y_i + kz_i - Y_i = ce^{\frac{1}{\Phi_i}\left(KG + \frac{1}{\psi_i}\right)\xi}. \tag{77}$$

Solving (75) gives

$$K_1 = \frac{1}{2}\left(\Phi_i - \frac{1}{G\psi_i}\right) + \sqrt{\frac{1}{4}\left(\Phi_i - \frac{1}{G\psi_i}\right)^2 + \frac{\Phi_i}{G}} \quad (78)$$

$$K_2 = \frac{1}{2}\left(\Phi_i - \frac{1}{G\psi_i}\right) - \sqrt{\frac{1}{4}\left(\Phi_i - \frac{1}{G\psi_i}\right)^2 + \frac{\Phi_i}{G}}. \quad (79)$$

From (77) follows that

$$\begin{aligned} y_i + K_1 z_i - Y_i &= c_2 e^{\frac{1}{\Phi_i}\left(K_1 G + \frac{1}{\psi_i}\right)\xi} \\ y_i + K_2 z_i - Y_i &= c_3 e^{\frac{1}{\Phi_i}\left(K_2 G + \frac{1}{\psi_i}\right)\xi}. \end{aligned} \quad (80)$$

Thus one obtains a system of three algebraic equations with three unknowns

$$y_i - \frac{1}{G} z_i - \Psi_i Y_i = c_1 \quad (81)$$

$$\begin{aligned} y - k_1 z_i - Y_i &= c_2 e^{\frac{1}{\Phi_i}\left(K_1 G + \frac{1}{\psi_i}\right)\xi} \\ y_i - k_2 z_i - Y_i &= c_3 e^{\frac{1}{\Phi_i}\left(K_2 G + \frac{1}{\psi_i}\right)\xi}. \end{aligned} \quad (82)$$

Calculating (82) for y_i, z_i, and Y_i one gets

$$\begin{aligned} y_i &= \frac{1}{1-\psi_i}\left\{c_1 + \frac{1 + G\psi_i K_2}{G(K_1 - K_2)} c_2 e^{\frac{1}{\Phi_i}\left(K_1 G + \frac{1}{\psi_i}\right)\xi} - \frac{1 + G\psi_i K_1}{G(K_1 - K_2)} c_3 e^{\frac{1}{\Phi_i}\left(K_2 G + \frac{1}{\psi_i}\right)\xi}\right\} \\ z_i &= \frac{1}{K_1 - K_2}\left\{c_2 e^{\frac{1}{\Phi_i}\left(K_1 G + \frac{1}{\psi_i}\right)\xi} - c_3 e^{\frac{1}{\Phi_i}\left(K_2 G + \frac{1}{\psi_i}\right)\xi}\right\} \\ Y_i &= \frac{1}{1-\psi_i}\left\{c_1 + \frac{1 + GK_2}{G(K_1 - K_2)} c_2 e^{\frac{1}{\Phi_i}\left(K_1 G + \frac{1}{\psi_i}\right)\xi} - \frac{1 + GK_1}{G(K_1 - K_2)} c_3 e^{\frac{1}{\Phi_i}\left(K_2 G + \frac{1}{\psi_i}\right)\xi}\right\}. \end{aligned} \quad (83)$$

Thus (83) is a general solution of system (71). Partial solution after substitution of boundary conditions (69), (70) reads:

$$y_i = \varkappa\left[1 + \frac{1 + G\Psi_1 K_2}{G(K_1 - K_2)} e^{\frac{1}{\Phi_i}\left(K_1 G + \frac{1}{\psi_i}\right)(\xi - L)} - \frac{1 - G\psi_i K_1}{G(K_1 - K_2)} e^{\frac{1}{\Phi_i}\left(K_1 G + \frac{1}{\psi_i}\right)(\xi - L)}\right] \quad (84)$$

$$Y_i = \varkappa\left[1 + \frac{1 + GK_2}{G(K_1 - K_2)} e^{\frac{1}{\Phi_i}\left(K_1 G + \frac{1}{\psi_i}\right)(\xi - L)} - \frac{1 + GK_1}{G(K_1 - K_2)} e^{\frac{1}{\Phi_i}\left(K_2 G + \frac{1}{\psi_i}\right)(\xi - L)}\right] \quad (85)$$

where

$$\varkappa = \frac{1}{1 + \psi_i \dfrac{1 + GK_2}{G(K_1 - K_2)} e^{-\frac{1}{\Phi_i}\left(GK_1 + \frac{1}{\psi_i}\right)L} - \psi_i \dfrac{1 + GK_1}{G(K_1 - K_2)} e^{-\frac{1}{\Phi_i}\left(K_2 G + \frac{1}{\psi_i}\right)L}}.$$

In case of the non-linear isotherm numerical solution is looked for with the method of successive approximations. Further on it is assumed that the first component is preferentially adsorbed from a mixture of components. It is reasonable to integrate system (68) between L and 0 at a given value of the relative breakthrough concentration y_{in}. By analogy to the pure component adsorption one can use as a first approximation the solution of this problem at $\xi = L$ for the linear isotherm (eq. 84) and with the value y_{in} one can solve system (84) by the Runge-Kutta method carrying out integration between L and 0. At every integration step it is necessary to check the satisfiability of the conditions:

$$Y_1 > 0; \qquad y_1 < 1; \qquad Y_1 < 1. \tag{86}$$

If $y_{1n}^{(k)}$ considerably exceeds the true value of $y_{1n}^{(1)}$ one can get $y_{1n}^{(k)} > 1$ in the process of calculation. This can lead to $Y_{1n}^{(k)} > 1$ or $Y_{1n}^{(k)} < 0$ which deprives the problem of its physcial sense. In this case one should move on to calculations with a lesser value $y_{in}^{(k)}$. Inequalities (86) being satisfied and integration completed, $y_1(0)$ defined from (68) must be compared with $y_1(0)$ from (69).

$$\Delta_1 = y_1(0) - y_1^T = y_1(0) - 1 - \frac{1}{G} z_1(0). \tag{87}$$

The possibility to stop iterations is checked by comparing $y_{1n}^{(k)}$ with $y_{1n}^{(k-1)}$ through the formula

$$w_1 = \frac{y_{1n}^{(k)} - y_{1n}^{(k-1)}}{y_{1n}^{(k)}}. \tag{88}$$

In case $\Delta_1 > 0$ and $|w_1| > \varepsilon$ where ε — is the calculation accuracy, calculation should be carried out with lesser $y_{1n}^{(k)}$. If $\Delta_1 < 0$; $|w| > \varepsilon$ the calculation is to be done with greater values of $y_{1n}^{(k)}$. With $|w| \leqq \varepsilon$ the calculation for the first component should be stopped and one should turn over to the calculation for the second component etc. For component i the whole sequence of the above inequalities and the conditions

$$\Delta y_i = y_i(0) - y_i^{(T)}(0) = y_i(0) - 1 - \frac{1}{G} z_i(0) \tag{89}$$

$$|w_i| = \left| \frac{y_{in}^{(m)} - y_{in}}{y_{in}^m} \right| \leqq \varepsilon \tag{90}$$

are checked. When $\Delta_i > 0$ and $|w| > \varepsilon$ the calculation is carried out with greater $y_{in}^{(m)}$ until the condition

$$|w_n| \leqq \varepsilon \tag{91}$$

is satisfied for component n. This method makes it possible to solve the problems of the dynamic adsorption for any isotherm equation in analytical or tabular form. In case of the dynamic adsorption from mixtures the adsorption isotherm was described by the equation of the LANGMUIR type:

$$\varphi_i(Y_1, Y_2, \ldots, Y_n) = P_i Y_i \frac{\prod_{k=1}^{n_{1i}} (1 + P_k)}{\prod_{k=1}^{n} (1 + P_k) + \sum_{k=1}^{n} (1 + P_k) \prod_{j=1}^{n_{1k}} (1 + P_j)} \tag{92}$$

where

$$P_i = \frac{A_{i\infty}}{\Gamma_i c_{i0}}.$$

The constants were found by the method of least squares with the help of a digital computer.

The systems of mixtures of zeolite NaX plus multicomponent hydrocarbon and active carbon plus binary hydrocarbon were used as the objects for modelling. The investigation results showed that at the equal total concentration

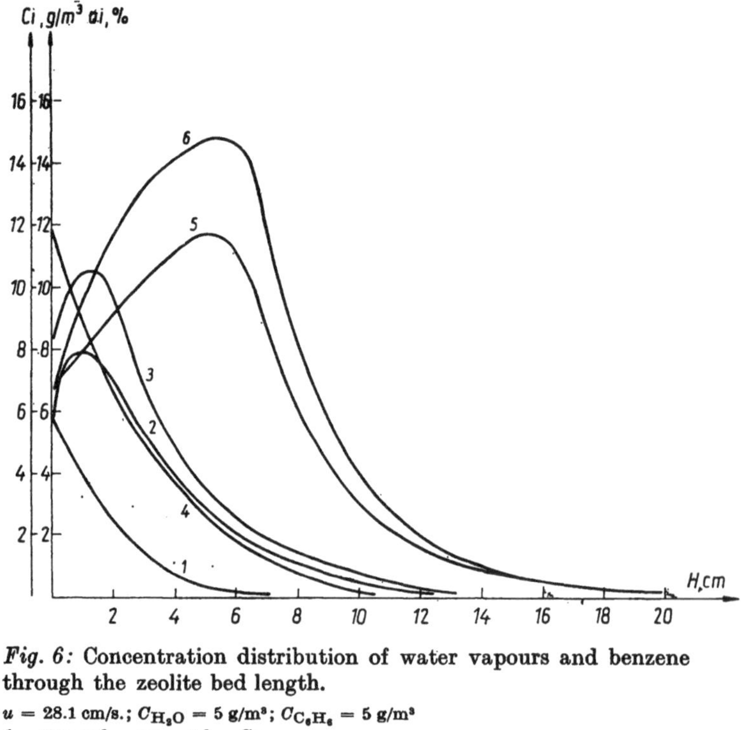

Fig. 6: Concentration distribution of water vapours and benzene through the zeolite bed length.

$u = 28.1$ cm/s.; $C_{H_2O} = 5$ g/m³; $C_{C_6H_6} = 5$ g/m³
1 – a_{H_2O}; *2* – $a_{C_6H_6}$; *3* – $C_{C_6H_6}$.
$C_{H_2O} = 10$ g/m³, $C_{C_6H_6} = 5$ g/m³ *4* – a_{H_2O};
5 – $C_{C_6H_6}$; *6* – $a_{C_6H_6}$;

of components the ratio of individual component concentrations is of substantial importance. At low concentrations of the more strongly adsorbed component preferential adsorption of the latter in the frontal beds is observed. It results in an increase of the mass-transfer zone length. Fig. 6 shows the results of digital computer calculation for the NaX$-H_2O-C_6H_6$ system. At a concentration of water and benzene vapour 5 mg/l the former is preferentially adsorbed and benzene is partially displaced from the frontal beds, its sconcentration increasing (curves 3.4).

When the concentration of water vapour rises up to 10 mg/l the frontal beds become saturated with water to 70—80%, the further concentrating of benzene continues (a_i and c_i increase) and the concentration maximum of benzene shifts to the right for 5—6 cm along the bed length. The total bed length where concentration drop is observed for both components is 20—22 cm. From the results of modelling follows that with great adsorbent flow rates (the amount adsorbed is small) distribution curves have the same form as those for pure component adsorption [18]. At the small flow rates of the adsorbent the amount adsorbed is great and considerable concentrating and displacement of the poorly adsorbed component is observed. Fig. 7 clearly shows the process of displacing the less strongly adsorbed benzene by water vapours. Thus, when

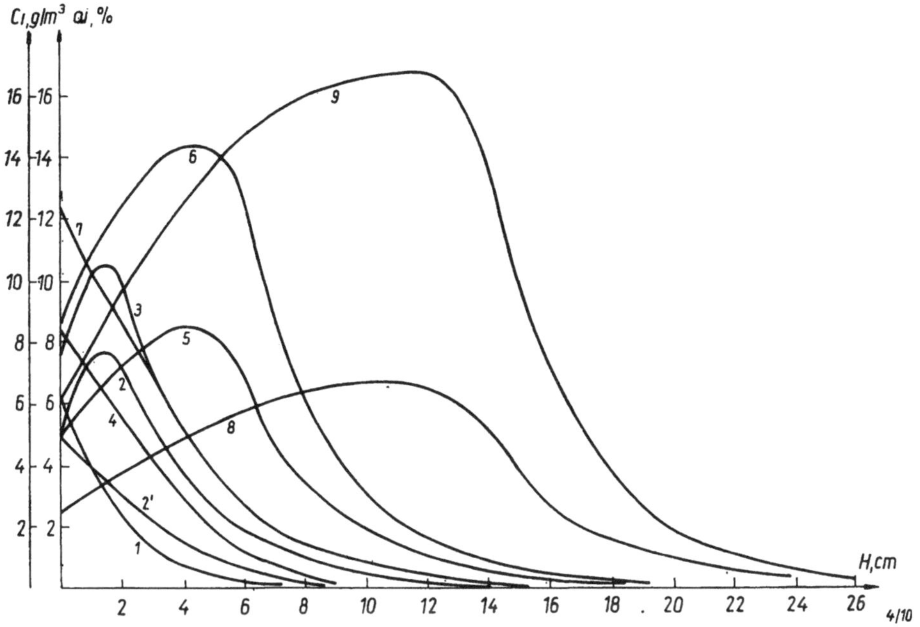

Fig. 7: Concentration distribution of water vapours and benzene through the zeolithe NaX bed length.

u = 28.1 cm/s, w = 15.7 cm/h, $C_{H_2O} = C_{C_6H_6}$ = 5 g/m³.
w = 11.8 cm/h: *4* — a_{H_2O}, *5* — $C_{C_6H_6}$, *6* — $a_{C_6H_6}$.
w = 7.8 cm/h: *7* — a_{H_2O}, *8* — $C_{C_6H_6}$, *9* — $a_{C_6H_6}$.

the zeolite flow rate drops from 15.7 to 7.8 cm/h NaX the bed length, where the concentration of benzene curves 3, 6, 9 and water vapours (1, 4, 7) decreases to a minimum fixed value, increases and is 15, 19, 28 and 6, 11, 15 cm, respectively.

Owing to the fact that benzene is adsorbed on the adsorbent preliminarily saturated with water the distribution curves for benzene concentration in gas and solid phase are considerably distorted and stretched through the total bed length. Due to the benzene concentration in the bed length of 8—16 cm the adsorption volume fraction of zeolite filled with benzene is 0.99. Fig. 8 shows the calculated distribution curves for the binary mixtures water + benzene and water + n-octane, differing in adsorbability of the second component. Since benzene is more strongly adsorbed and its displacement is more difficult than that of n-octane, the distribution curve for water (5) is more diffused and distorted which results in shifting the concentration maximum of benzene to the right. In spite of the fact that the mass transfer coefficient of water vapour was the same in calculations ($\beta_{H_2O} = 20\ s^{-1}$), its effect on the form of the distribution curves of benzene and n-octane was quite different. The figure shows the steeper leading front for n-octane, although $\beta = 5\ s^{-1}$ for the latter, and the steeper distribution curves of benzene at the back front, for benzene $\beta = 10\ s^{-1}$.

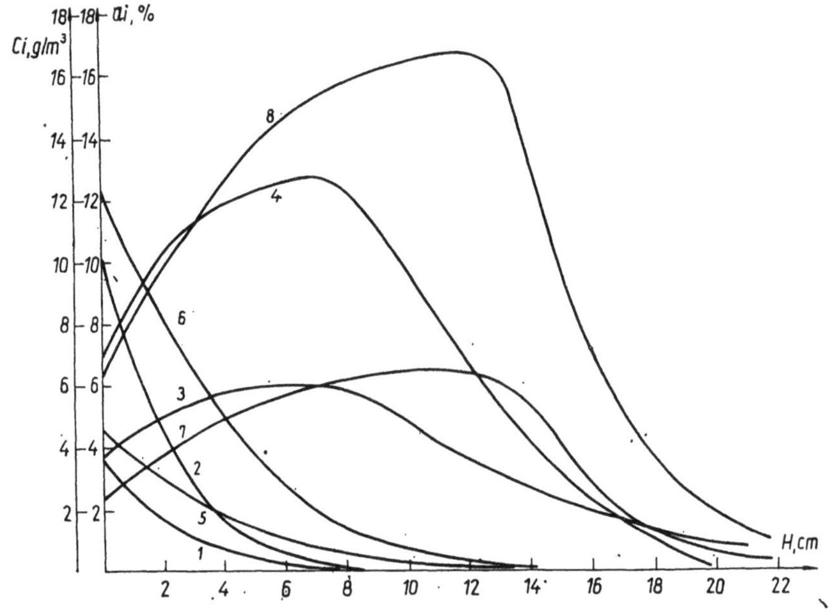

Fig. 8: Concentration distributions of water, benzene and n-octane through the zeolite NaX bed length.

$C_{H_2O} = C_{C_6H_6} = C_8H_{18} = 5$ g/m³, $w = 7.8$ cm/h;
$u = 28.1$ cm/s: H_2O — C_8H_{18}: *1* — C_{H_2O}, *2* — a_{H_2O}, *4* — $C_{C_8H_{18}}$.
H_2O — C_6H_6: *5* — C_{H_2O}, *6* — a_{H_2O}, *7* — $C_{C_8H_{18}}$, *8* — $a_{8H_{18}}$

Analysis of the kinetic characteristics (Fig. 9) showed that decrease of the kinetic constant of the first preferentially adsorbed component resulted in significant lengthening of the zone, where the largest concentration drop was noticed. This in turn was followed by the lengthening of the adsorption zone of the 2-nd less strongly adsorbed component. Furthermore the rise of the mass transfer coefficient for the I-st component made the leading adsorption front steeper but did not change the form of the back front.

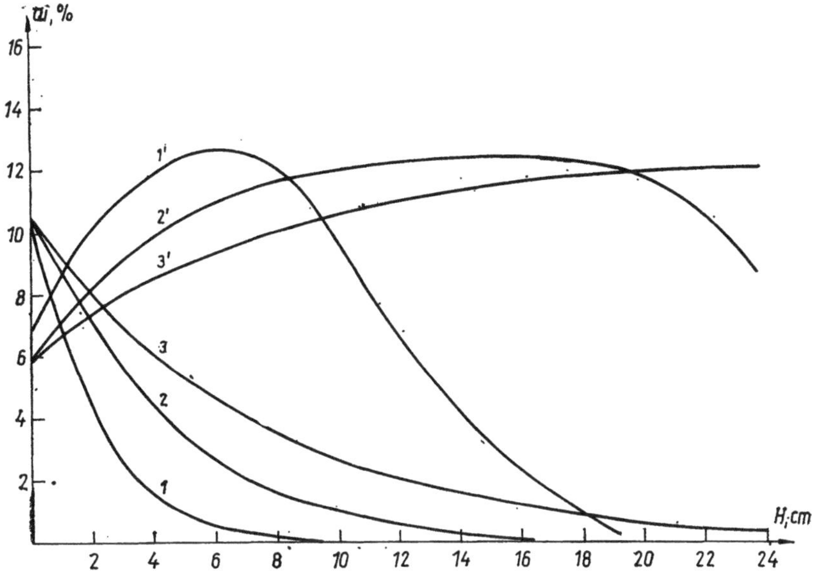

Fig. 9: Concentration distributions through bed length.
$C_1 = C_2 = 5$ g/cm³, $w = 7.8$ cm/h, $u = 28.1$ cm/s.
$\beta_1 = 20$, $\beta_2 = 5$: 1 – Y_1, 1' – Y_2.
$\beta_1 = 10$, $\beta_2 = 5$: 2 – Y_1, 2' – Y_2.
$\beta_1 = 5$, $\beta_2 = 5$: 3 – Y_1, 3' – Y_2.

Fig. 10, 11 show calculation and experimental results for the adsorption of a three-component-mixture on zeolite NaX. Even a small drop of the adsorbent flow rate leads to redistribution of the gas and solid phase concentrations and changing of the distribution curve shape. In our calculations mean integral values of mass transfer coefficients are used and the adsorption isotherm is defined by the LANGMUIR equation. Good agreement is found between the theory and experimental results. The calculations are carried out with individual component characteristics and therefore may be fulfilled for different initial concentrations and adsorbent flow rates. It is clear that calculation with digital computers excludes the need of tedious and time consuming experiments and makes possible evaluation of solid and gas phase concentrations and concentration maxima for any given operating conditions in adsorption columns. The calculation results may be used for the optimization of the operating conditions and dimensioning of the adsorption apparatus.

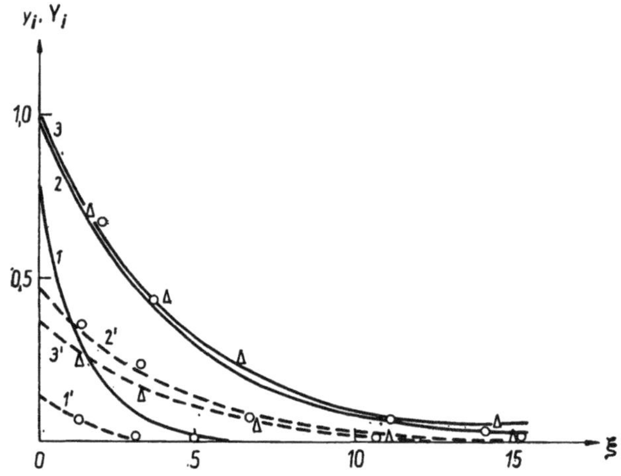

Fig. 10: Concentration distributions through zeolite bed length.

1 – y_{H_2O}, 2 – $y_{C_6H_6}$, 3 – $y_{C_8H_{18}}$
1' – Y_{H_2O}, 2' – $Y_{C_6H_6}$, 3' – $Y_{C_8H_{18}}$.
$u = 28.1$ cm/s, $w = 14.5$ cm/h,
$C_{H_2O} = 1.4$ g/m³, $C_{C_6H_6} = 4.1$ g/m³, $C_{C_8H_{18}} = 2.8$ g/m³.

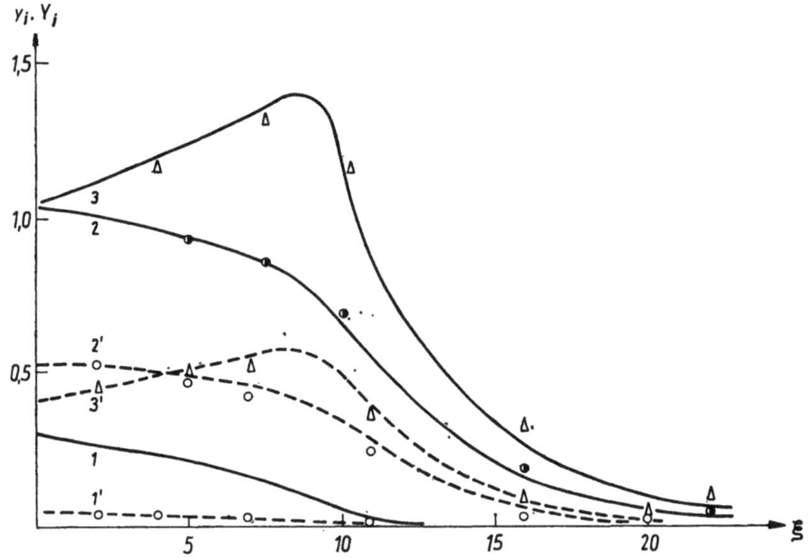

Fig. 11: Concentration distributions through zeolite NaX bed length.

$u = 28.1$ cm/s, $w = 12.5$ cm/h,
$C_{H_2O} = 1.4$; $C_{C_6H_6} = 4.1$; $C_{C_8H_{18}} = 2.8$ g/m³.

In the description of the process of a moving bed by the displacement technique adsorbent flow uniformly saturated with a desorbed component is fed from the top of an adsorption column at the flow rate w. From the bottom of the column ($x = 0$) gasvapour mixture containing the concentration of a desorbing agent, c_{01}, is fed at the flow rate u. In this case the problem reduces itself to the definition of the adsorbent bed length H, in which the key component concentration reaches a predetermined value. The displacement desorption process is defined by the system of differential equations developed for the mixed adsorption case:

$$u \frac{dc_i}{dx} = -\beta_i[c_i - \varphi_i(a_1, a_2)]$$

$$-w \frac{da_i}{dx} = \beta_i[c_1 - \varphi_i(a_1, a_2)]. \tag{93}$$

The boundary conditions are:

$$x = 0 \qquad c_1 = c_{01} \qquad c_2 = 0 \tag{94}$$

$$x = H \qquad a_1 = 0 \qquad a_2 = a_{02}. \tag{95}$$

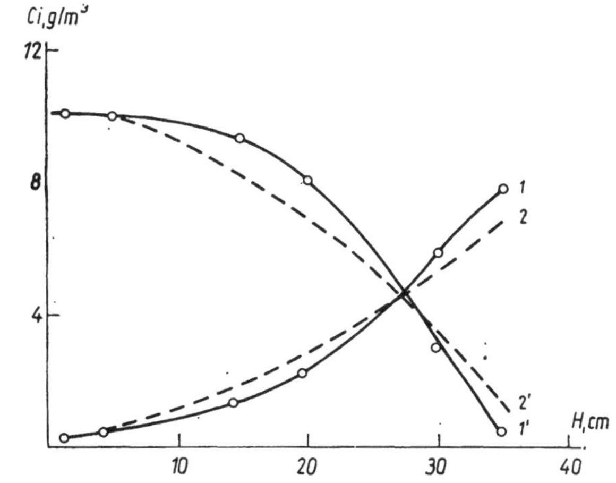

Fig. 12: Concentration distributions through zeolite NaX bed length.
(1–2) – C_8H_{18}, (1'–2') – C_6H_6
w = 223 g/h, $C_{C_6H_6}$ = 10.4 g/m³,
Experimental data are indicated by dots.

Fig. 12 shows the calculated and experimental curves of n-octane desorption by benzene from zeolite NaX. Some discrepancy between the theory and experiment is probably caused by inaccurate determination of the time, necessary for steady-state process formation under laboratory conditions and of the component concentrations, which differ by several orders. Since the working bed length in the continuously operating adsorption columns with moving beds is rather small, high productive efficiency of the process is provided by the apparatus with the small ratio.

This is inconvenient from the designing point of view and may be overcome by the use of phase crossflow according the scheme shown in Fig. 13. The other and more effective way is to use the crossflow of phases according to the scheme shown in Fig. 13. In this case the mathematical model of the process must be written in the following form:

$$\frac{dc}{dt} = \text{div } Dc \text{ grad } c - \text{div } (uc) - G_1$$
$$\frac{da}{dt} = \text{div } Da \text{ grad } a - \text{div } (wa) + G_1 \tag{96}$$

where Dc and Da — the coefficients characterising the component mass transfer in the corresponding phase and G is the source function.

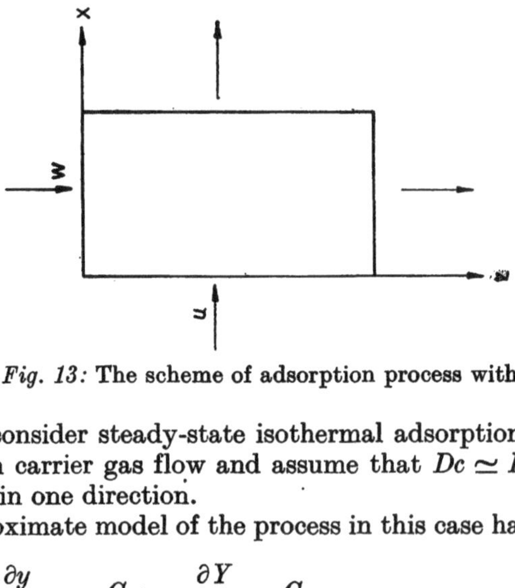

Fig. 13: The scheme of adsorption process with phase crossflow.

Now we consider steady-state isothermal adsorption of an individual component from carrier gas flow and assume that $Dc \simeq Da = 0$ and each phase moves only in one direction.

The approximate model of the process in this case has the following form:

$$\frac{\partial y}{\partial \xi} = -G_1; \quad \frac{\partial Y}{\partial \tau} = G_1$$
$$\xi = \frac{x\beta}{u}; \quad \tau = \frac{\beta \cdot z}{\Gamma \cdot w} \tag{97}$$

where G the dimensionless form of the kinetic equation. Thus, the initial and boundary conditions for the process in Fig. 13 can be written in the form:

$$y(0, \tau) = 1; \quad Y(\xi, 0) = Y_H.$$

Now it is evident that the analysed situation is similar to the frontal chromatography problem when axial dispersion effects are negligibly small. Solution of such problems were obtained for different G-functions.

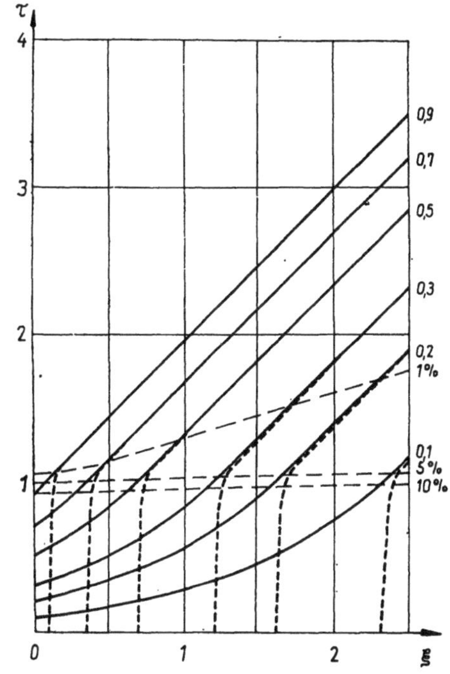

Fig. 14: Adsorption process diagramme with kinetics limited by external diffusion.

$P = 0.25$ – – – $y = \text{const.}$; $Y = \text{const.}$

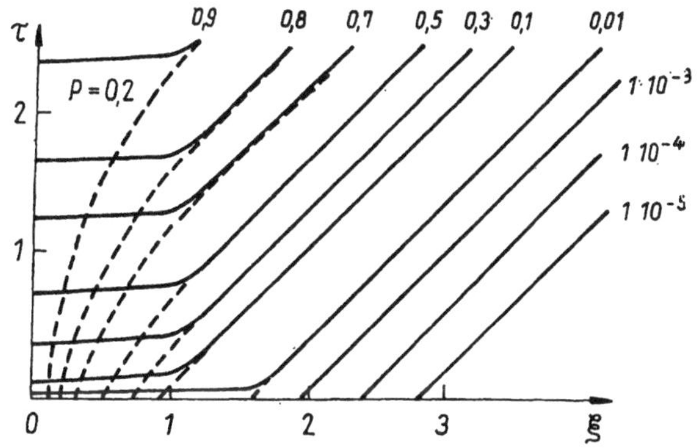

Fig. 15: Adsorption process dynamics with the kinetics equation proposed by GLUECKAUF.

$P = 0.25$

– – – $y = \text{const.}$; $Y = \text{const.}$

Fig. 14 gives an example of adsorption process development described by model (97) where $G_1 = y_1 - \frac{PY}{1+p-Y}$ i.e. for the LANGMUIR type isotherm and adsorption kinetics limited by external diffusion. Fig. 15 shows the same when $G_1 = y\,\frac{p+1}{p+y} - Y$ i.e. for the LANGMUIR isotherm and the kinetic equation proposed by Glueckauf.

In distinction from ordinary frontal chromatography in this case the flow eluating from the apparatus contains the concentration of the adsorbed component:

$$y_k = \frac{1}{\tau_n}\int_0^{\tau_n} y(H, \tau)\, d\tau. \tag{98}$$

If the crossflow adsorption process is carried out in several steps it may be calculated with a digital computer by the method of successive approximations using model (96) and a corresponding kinetic equation or simple asymptotic approximations at sufficiently high values of H.

5. References

[1] BERG, C., Chem. Eng. Progr. **47**, 11, 585 (1951)
[2] BERG, C., Sci. Petrol. **5**, 2, 142 (1953)
[3] HARRIS, G., Chem and Process Eng. **47**, 8, 45 (1966)
[4] ROMANKOV, P. G., LEPILIN, V. N., "*Continuous Adsorption of Gases and Vapours*", Chemistry, Leningrad, 1968
[5] SOKOLOV, V. A., "*New Methods of Separation of Light Hydrocarbons*", Gostoptechisdat, Moscow 1961
[6] KELCEV, N. V., "*Basic Principles of Adsorption*" Chemistry Moscow, 1976
[7] KOUL, A. L., RIESENFELD, F. S., Gas Purification, Gostoptechnisdat Moscow 1962
[8] BENEDEK, P., SEPESCHI, L., Nad. Z. Gas. Industry **30** (1958)
[9] PLACHENOV, T. G., Com. Ac. Sc. USSR, Serie Chemistry, **6**, 968 (1965)
[10] TODES, O. M., "*Methods and Processes of Chemical Technology*", Academy, Moscow 1955
[11] ROMANKOV, P. G., RASHKOVSKAYA, N. B., FROLOW, N. F., "*Maso Transfer Processes in Chemical Technology*", Chemistry, Leningrad 1975
[12] KASTEN, P. R., AMUNDSON, N. R., Ind. Eng. Chem. 441, 1704 (1952)
[13] ROMANKOV, P. G., LEPILIN, V. N., FROLOV, V. F., "*Kinetics and Dynamics of Physical Adsorption*", Nauka, Moscow 1973
[14] SEBALLO, A. A., BARANOV, E. I., LESIN, Ju. S., PLACHENOV, T. G., "*Kinetics and Dynamics of Physical Adsorption*", Nauka, Moscow 1973
[15] PLACHENOV, T. G., LESIN, Ju. S., BARANOV, E. I., SEBALLO, A. A., SHIRYAEV, A. N., Teor. Found. Chem. Techn. **6**, 4, 622 (1972)
[16] VETROV, B. N., TODES, O. M., Journal Techn. Phys. **25**, 7, 1217 (1955)
[17] BARANOV, E. I., LESIN, Ju. S., PLACHENOV, T. G., SEBALLO, A. A., SHIRYAEV, A. N., "*Kinetics and Dynamics of Physical Adsorption*", Nauka, Moscow 1973
[18] PLACHENOV, T. G., SEBALLO, A. A., LESIN, Ju. S., BARANOV, E. I., KWASCHA, V. I., Journal Appl. Chemistry **50**, 9, 1994 (1976)

On the mathematical simulation of mass-transfer processes in membrane systems

(I. N. Taganov, V. I. Tcheremisin

Zusammenfassung:

Gegenstand des Beitrages sind Untersuchungen an Hand mathematischer Modelle von Stofftransportprozessen durch Membranen, die hohe Selektivität besitzen. Die vorgeschlagenen Modelle berücksichtigen den Einfluß nichtlinearer Effekte, die mit der Sättigung der Membranen und den chemischen Wechselwirkungen des Mediums mit der Membranstruktur verbunden sind.

Ein spezielles mathematisches Modell des Diffusionsstromes, das nichtlinear bezüglich der Konzentrationsgradienten des transportierten Stoffen ist, wird abgeleitet. Bei linearen Gradienten geht es in das 1. Ficksche Gesetz über. An Hand dieses Modells werden Differentialgleichungen für den Stofftransport an Membranen bei Diffusion und Filtrationsmechanismus abgeleitet und untersucht. Quellen in Form von chemischen Reaktionen werden berücksichtigt.

1. Introduction

In recent years interest in the processes of mass, heat and charge transfer in membrane systems has increased greatly. To a great extent this is connected with the attempts of practical utilization of biological membrane functioning principles, the most important of which is their capability to accomplish a selective exchange of molecules of various substances. It was this capability that lately led to the development and production of artificial membrane materials with selective properties. On the other hand usage of artificial semipermeable selective membranes results in developing a new field in chemical technology — membrane technology.

While designing mathematical models of processes of transfer through the membranes investigators face many problems.

Primarily the difficulties arise from the fact that a throughput mechanism of both biological and artificial membrane materials has not been investigated completely and a general approach related to it is missing.

There are two main approaches to the membrane transport investigation. The thermodynamical approach gives a phenomenological description of the transfer process from the standpoint of irreversible process thermodynamics [1—3]. The second approach is to create a mathematical model, in which the membrane is considered as a homogeneous phase or as a succession of potential barriers [4—5].

The development of a versatile model adequately describing membrane transport is complicated by a variety of membrane tranfer processes. In biological membrane systems reference should be made to the following types of process — passive transport (diffusion of general type), active transport (mass transfer against the gradient of concentration) and lightened diffusion (mass transfer along the gradient with an anomalously high rate).

As for artificial membranes used in chemical technology mention should be made of filtration and diffusion models, which are in common use. The former presents the membrane as a capillary-porous medium, in which mass transfer is effected by filtration flows [6, 7], for the latter is based on the diffusion mechanism of penetration [6, 8—10]. To describe the mass transfer phenomenon through biological membranes both common diffusion and filtration probability models were used [11]. The mathematical description of active transport and lightened diffusion is more complicated. Here besides the discrete models mentioned above diffusion models with a carrier [12] and sorption-diffusion models [13, 14] are employed.

Diffusion models commonly encountered in special literature treat the relation between transferred mass concentration gradient and mass flow according to conventional linear form of FICK's law.

Diffusion coefficient in so doing is assumed to be constant and membrane material is not taken into consideration. The membrane transport process singularity results from the presence of transmembrane gradients of concentration, temperature, pressure, electric potential and is specified by the membrane material structure and its interaction with transferred material molecules.

In increasing the transmembrane concentration gradient as follows from FICK's 1st law mass flow increases linearily and should reach a high value with significantly high gradient values. Practically this condition is not fulfilled and linear dependence holds only for low gradients and comparatively thick membranes.

The present paper is devoted to the investigation of non-linear models for flows reaching saturation with a large concentration gradient.

2. Transfer equation and non-linear mass flows

According to [15] the mass transfer equation for a moving compressed medium may be written in the following way:

$$\frac{dC}{dt} + \operatorname{div} C\bar{v} = -\operatorname{div} \bar{q} + I \qquad (1)$$

where:

C — volumetric concentration of the mass;
$\bar{v}$ — rate vector;
$\bar{q}$ — diffusion flow through unitary surface — UMOV's vector;
I — volumetric power of mass sources (or drainage)
t — time

Equation (1) can be written in the following way:

$$\frac{dC}{dt} = -\text{div}\,(\bar{q} + \bar{v}C) + I \tag{2}$$

from which, taking into consideration $\text{div}\,(\bar{v}C) = C\,\text{div}\,\bar{v} + \bar{v}\,\text{grad}\,C$, the relationship equation for non-compressible liquid can be derived:

$$\frac{dC}{dt} = -\text{div}\,\bar{q} - \bar{v}\,\text{grad}\,C + I. \tag{3}$$

For an n-component system we have a system of n equations of 2 or 3 types. In this case the accuracy of theoretical description depends on the interaction of diffusion flows $\bar{q}$, rate field $\bar{v}$ and mass source (or drainage) functions I resulting from phase or chemical transformation.

Let us confine ourselves to the consideration of single-component systems. The diffusion mass flow in the conventional form of FICK's 1st law is:

$$\bar{q} = -D\,\text{grad}\,C \tag{4}$$

where in general D is a tensor of the second order with components depending on co-ordinates and concentration, in the case of an isotropic medium the tensor being diagonal and in so doing

$$D_{ii} = D \tag{5}$$

where D — diffusion coefficient.

For the isotropic medium the tensor D can be converted to diagonal form by selecting a coordinate system. The value of $\bar{q}$ flow can be calculated from the following relationship:

$$\|\bar{q}\| = \sqrt{(\bar{q}, \bar{q})} \tag{6}$$

where $(\bar{q}, \bar{q})$ is a scalar product. (4) and (6) give

$$\|\bar{q}\| = D\,\|\text{grad}\,C\| \tag{7}$$

for an isotropic medium.

Our model is based on the assumption that with high concentration gradients the diffusion flow is limited by a certain limiting value of $\bar{q}_s$ saturation. Thus flow $\bar{q}$ is a non-linear fuction of the concentration gradient or a linear function of a certain concentration quasi-gradient, which can be expressed in the following way:

$$Q\,\text{grad}\,C = f(\|\text{grad}\,C\|)\,\text{grad}\,C. \tag{8}$$

As in an anisotropic medium various limitations of grad C vector components can exist, in general f is the 2nd order tensor. Thus for flow $\bar{q}$ we have:

$$\bar{q} = -D^*Q\,\text{grad}\,C \tag{9}$$

where D^* is an analogy of D in expression (4). For the membranes representing thin films and possessing a sufficiently anisotropic expression flow (9) is simplified greatly inasmuch as we have:

$$D_{11}^* = D^*, \quad D_{22}^* = D_{33}^* = 0; \quad f_{11} = f, \quad f_{22} = f_{33} = 0. \tag{10}$$

Therefore the problem turned into a one-dimensional one, and flow $\bar{q}$ is directed along the normal to the membrane surface. It should be noted that in the general case under constant temperature and pressure diffusion coefficient D^* is a function of coordinates and concentration:

$$D^* = D^*(x, C). \tag{11}$$

The question of the choice of the function f in the expression for quasi – gradient (8) in designing a concrete non-linear model for diffusion flow in a membrane arises above all. The authors suggest to take

$$f(\|\text{grad } C\|) = \frac{1}{\lambda \, \|\text{grad } C\|} \cdot (1 - e^{-\lambda \|\text{grad } C\|}) \tag{12}$$

as function f. In expression (12) parameter λ characterises the medium resistance to transfer.

As in the one-dimensional problem we have

$$\|\text{grad } C\| = |\text{grad } C| \equiv \left| \frac{dC}{dx} \right|,$$

a diffusion flow $\bar{q}$ through the membrane is:

$$\bar{q} = -q_s \frac{\text{grad } C}{|\text{grad } C|} (1 - e^{-\lambda |\text{grad } C|}), \quad q_s = \frac{D^*}{\lambda}. \tag{13}$$

Dependences between diffusion flow magnitude and concentration gradient for linear (4) and non-linear (13) flows give the following relationship:

a) linear flow $\quad |\bar{q}| = D \, |\text{grad } C|, \quad$ (14)

b) non-linear flow $\quad |\bar{q}| = q_s(1 - e^{-\lambda |\text{grad } C|}). \quad$ (15)

In Fig. 1 line a corresponds to linear flow and line b to nonlinear flow.

It should be noted that expression (12) for the non-linear diffusion flow with low concentration gradient reduces to the general FICK law. It can be substantiated by expanding the exponential in expression (12) in power series of $\lambda \times |\text{grad } C|$ and considering in so doing only the terms of the first order.

If a membrane represents a capillary-porous substance, besides the transfer diffusion mechanism there is a transfer filtration mechanism in it. Filtration rates are calculated from the equation of D'ARSY [8]:

$$\bar{v}_f = -\frac{k\varrho}{\eta} \text{grad } \varphi \tag{16}$$

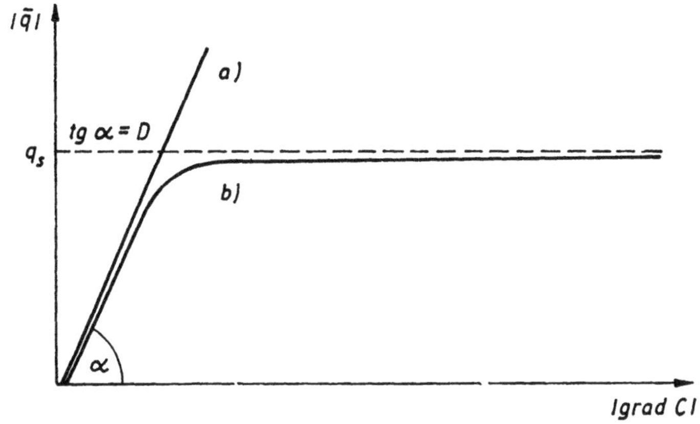

Fig. 1: The dependence of the value of diffusion flow on the concentration gradient.

a) linear flow.
b) nonlinear flow with the saturation.

where:

$\bar{v}_f$ — filtration volumetric rate;
k — permeability coefficient;
η — viscosity (dynamic coefficient of viscosity);
ϱ — liquid density;
φ — force potential estimated for the one-dimensional problem in the following way;

$$\varphi = gx + \int_{p_0}^{p} \frac{dp}{\varrho(p)} \tag{17}$$

where:

p_0 — pressure when $x = 0$
g — gravitational field.

If we assume that the filtration rate is a limited function of the pressure gradient, the filtration rate expression can be written in a form similar to (9).

3. Transfer equations for non-linear flows

The diffusion transfer of a mass [7] through the membrane consists of the following steps:

1. mass transfer from the solution washing the membrane to the membrane surface (convection diffusion);
2. mass solution in the membrane material;
3. diffusion transfer in the membrane material;
4. mass desorption on the back side of the membrane;
5. desorbed mass removal into the surrounding solution (diffusion and convection).

The filtration mechanism of transfer exists in the membranes with capillary-porous structure, the membrane selectivity depending upon many factors such as: the nature of the dissolved substance, size of the membrane pores, membrane porosity, trans-membrane pressure drop value, etc.

More complicated mechanisms, in which substance transfer is accomplished by both diffusion and filtration, requires the solution of a system of two non-linear transient equations, which is not considered in the present paper.

The transfer equation for the diffusion mechanism of penetration can be derived from (2) when $\bar{v}$ is assumed to be equal to 0, and substituting the expression for diffusion flow (13) into (2) gives:

$$\frac{\partial C}{\partial t} - \operatorname{div}\left[D_1^*(x, C)\,\frac{\operatorname{grad} C}{\vartheta}\,(1 - e^{-\vartheta})\right] \pm r(C) = 0 \tag{18}$$

where $\vartheta = \lambda\,|(\operatorname{grad} C)|$.

Kinetic term $r(C)$ reflects the interaction of the transferred mass and membrane material. Let us note that equation (18) is equivalent to two following equations:

$$\frac{\partial C}{\partial t} - \frac{\partial}{\partial x}\left[D_1^*(x_1C)\,\frac{1}{\lambda}\left(1 - e^{-\lambda\frac{dC}{dx}}\right)\right] \pm r(C) = 0, \quad \text{for} \quad \frac{dC}{dx} > 0 \tag{19}$$

$$\frac{\partial C}{\partial t} - \frac{\partial}{\partial x}\left[D_1(x_1C)\,\frac{1}{\lambda}\left(1 - e^{\lambda\frac{dC}{dx}}\right)\right] \pm r(C) = 0, \quad \text{for} \quad \frac{dC}{dx} < 0. \tag{20}$$

The following dependences for diffusion coefficient D are the most frequently used in literature [7, 8, 16]

$$D(C) = D_0 e^{\beta C} \tag{21}$$

$$D(C) = D_0(1 + \beta C) \tag{22}$$

$$D(x) = D_0\big(1 + f(x)\big). \tag{23}$$

Let us pass to the deriving of the transfer differential equation ofr the filtration mechanism of penetration. For simplicity let us confine ourselves to the consideration of incompressible liquid. Substituting the expression for diffusion flow (13) into the equation (3) gives:

$$\frac{\partial C}{\partial t} - \frac{\partial}{\partial x}\left[D_2^*(x, C)\,\frac{dC}{dx}\,(1 - e^{-\vartheta})\,\frac{1}{\vartheta}\right] + v(x, t)\,\frac{dC}{dx} \pm r(C) = 0, \tag{24}$$

filtration volumetric rate $v(x, t)$ depending upon trans-membrane pressure drop, size of pores, viscosity etc. D_2^* — diffusion coefficient of the substance dissolved in a liquid phase. Equation (24) as well as equation (18) can be transformed to two single equations for positive and negative concentration gradients.

4. Simplified non-linear models of membrane transport

The analysis of the transfer differential equations of general type derived on the base of the non-linear flows is rather complicated. Therefore let us simplify them. Practically interesting are the solutions for steady conditions. Assuming that $dC/dt = 0$ in (18), gives:

$$\frac{d}{dx}\left[D_1^*(x, C)\frac{dC}{dx}(1 - e^{-\vartheta})\frac{1}{\vartheta}\right] \mp r(C) = 0. \tag{25}$$

Equation (25) is transformed to

$$\frac{d^2C}{dx^2} D_{10}^* e^{-\vartheta} \mp r(C) = 0 \tag{26}$$

for constant diffusion coefficient. Equation (26) transforms to

$$\frac{d}{dx}\left[D_1^*(x, C)\frac{dC}{dx}(1 - e^{-\vartheta})\frac{1}{\vartheta}\right] = 0 \tag{27}$$

when there is no chemical interaction between transferred mass and membrane material, the diffusion coefficient depending upon coordinate x and concentration.

The diffusion flow, as follows from equation (27), does not depend on the coordinate:

$$\frac{dC}{dx}(1 - e^{-\vartheta})\frac{1}{\vartheta} = \frac{q_0^*}{D_1^*(x, C)}. \tag{28}$$

Finally assuming that the diffusion coefficient is constant ($D_1^* = \text{const.} = D_{10}^*$), and expecting the chemical interaction to be missing we get:

$$\frac{dC}{dx}(1 - e^{-\vartheta})\frac{1}{\vartheta} = \frac{q_0^*}{D_{10}^*}. \tag{29}$$

In case of steady conditions we get from (24):

$$\frac{d}{dx}\left[D_2^*(x, C)\frac{dC}{dx}(1 - e^{-\vartheta})\frac{1}{\vartheta}\right] - v(x)\frac{dC}{dx} \pm r(C) = 0 \tag{30}$$

for the filtration mechanism of penetration. If the diffusion coefficient ($D_{20}^* = \text{const.}$) and the rate of filtration ($v = \text{const.} = v_0$) are constant, it follows from (30) that:

$$\frac{d^2C}{dx^2} D_{20}^* e^{-\vartheta} - v_0\frac{dC}{dx} \mp r(C) = 0. \tag{31}$$

When the interaction between transferred substance and membrane material is missing, and the rate of filtration v_0 is constant, and $D_2^* = D_2^*(x, C)$, we have:

$$\frac{d}{dx}\left[D_2{}^*(x, C)\,\frac{dC}{dx}\,(1 - e^{-\vartheta})\,\frac{1}{\vartheta}\right] = v_0\,\frac{dC}{dx} \tag{32}$$

hence

$$D_2{}^*(x, C)\,\frac{dC}{dx}\,(1 - e^{-\vartheta})\,\frac{1}{\vartheta} = v_0 C + \text{const.} \tag{33}$$

Finally if the diffusion coefficient is $D_2{}^* = \text{const.} = D_{20}^*$, then

$$D_{20}^*\,\frac{dC}{dx}\,(1 - e^{-\vartheta})\,\frac{1}{\vartheta} = v_0 C + \text{const}. \tag{34}$$

5. Special types of transfer equations and their solutions

In general the mass transfer under steady conditions is represented by equation (25). If $D_1{}^*(x, C)$ is assumed to be the function of C only, i.e. $D_1{}^* = D^*(C)$, equation (25) transforms to:

$$\frac{d}{dx}\left[\frac{D^*(C)}{\lambda}\left(1 - e^{-\lambda\frac{dC}{dx}}\right)\right] - r(C) = 0 \tag{35}$$

for positive gradients.

The left part of the equation (35) does not involve x and its solutions $C = C(x)$ under $C' \neq 0$ have inverse functions $x = x(C)$. Therefore derivatives C'' and C' are certain functions of C. Taking into consideration $C' = \dfrac{dC}{dx} = p(C)$ gives:

$$\frac{d^2C}{dx^2} = \frac{dp(C)}{dC}\,\frac{dC}{dx} = p'(C)\,p(C). \tag{36}$$

Substituting (36) in equation (35) gives

$$p(C)\,\frac{d}{dC}\left[\frac{D^*(C)}{\lambda}\,(1 - e^{-\lambda p(C)})\right] - r(C) = 0. \tag{37}$$

Equation (38) can be transformed to:

$$\frac{d}{dC}\left[\frac{p(C)\,D^*(C)}{\lambda}\,(1 - e^{-\lambda p(C)})\right] = r(C) + \frac{D^*(C)}{\lambda^2}\,\frac{d}{dC}\left(p(C)\,\lambda + e^{-\lambda p(C)}\right) \tag{38}$$

integration of which when $r(C) = kC^n$ and $D^*(C) = \text{const.} = D_0{}^*$ gives a non-linear equation for the determination of $p(C)$ as a function of C:

$$-e^{-\lambda p(C)}\left(1 + \lambda p(C)\right) = \frac{\lambda^2}{D_0{}^*}\left[\frac{kC^{n+1}}{n+1} + A_0\right] = f(C) \tag{39}$$

where A_0 — integration constant.

When $\lambda p(C) = -\ln \mu$ is assumed, then equation (39) is transformed to:

$$\mu(\ln \mu - 1) = f(C). \tag{40}$$

A transcendental equation (40) can be approximately solved against μ. The concentration distribution in the membrane for approximate dependance $\mu = \mu(C)$ is derived from:

$$\frac{dC}{dx} = -\frac{1}{\lambda} \ln \mu(C) \tag{41}$$

for which only an approximate numerical or analytical solution can be obtained. A precise analytical solution of equation (35) can be obtained when $D^*(C) = \text{const.} = D_0^*$ and when the reaction of zero order takes place $r(c) = k$. In so doing equation (36) gives:

$$\frac{d}{dx} \left[\left(1 - e^{-\lambda \frac{dC}{dx}}\right)\right] = \frac{\lambda k}{D_0^*}. \tag{42}$$

Integrating (42), and taking logarithm, gives an equation:

$$\frac{dC}{dx} = -\frac{1}{\lambda} \ln \left[1 - A_0 - \frac{\lambda k}{D_0^*}\right] x. \tag{43}$$

Equation (42) implies that the flow linearily depends upon coordinate x. Integrating (43) gives concentration distribution along the membrane thickness:

$$C(x) = -\frac{1}{\lambda} \left[\left(\ln \frac{\lambda k}{D_0^*} - 1\right) x - (a - x) \ln (a - x) + B_0\right], \tag{44}$$

where

$$a = \frac{D_0^*(1 - A_0)}{\lambda k}. \tag{45}$$

Integration A_0 and B_0 are defined from boundary conditions of the membrane.

Let us consider the case of non-linear mass diffusion in which mass with the membrane material is missing. For simplicity let us assume that diffusion coefficient D^* is a function of concentration only i.e. it does not depend evidently on coordinate x. The steady condition equation in this case is:

$$\frac{d}{dx} \left[D^*(C) \frac{dC}{dx} \frac{1}{\vartheta} (1 - e^{-\vartheta})\right] = 0 \tag{46}$$

and for positive gradients:

$$\frac{D^*(C)}{\lambda} \left(1 - e^{-\lambda \frac{dC}{dx}}\right) = \text{const.} = q_0 \tag{47}$$

i.e. the flow does not depend on the coordinate. The non-linear equation for the concentration profile in the membrane is derived from (47):

$$\frac{dC}{dx} = -\frac{1}{\lambda} \ln \left[1 - \frac{q_0}{D(C)}\right] = -\frac{1}{\lambda} f(C). \tag{48}$$

Equation (48) is an equation with separable variables and its integration gives:

$$\int_{C_0}^{C} \frac{dC}{f(C)} = -\frac{x}{\lambda} + A_0. \tag{49}$$

For every concrete $D^*(C)$ the solution acquires $x = x(C)$ form and the integration constants q_0 and A_0 are defined from boundary conditions. The integral in the left part of equation (49) when $D(C) \neq$ const. is taken approximately. Dependence $C = C(x)$ can be defined approximately either numerically or analytically, reverse interpolation being used in the numerical solution. For inear mass flow FICK's 2nd law implies that:

$$\frac{1}{q_0^*} \int_{C_0}^{C} D^*(C)\, dC = x + A_0^* \tag{50}$$

in which q_0^* and A_0^* — values of the flow for the linear model and integration constant, respectively.

An approximate analytical solution of equation (48) can be obtained by expanding the integrated function in (49) in power series of $q_0\lambda/D^*(C)$:

$$-\frac{1}{\lambda} \ln \left[1 - \frac{q_0}{D^*(c)(C)}\right] = \frac{q_0}{D^*(C)} \left[1 + \frac{1}{2} \frac{q_0\lambda}{D^*(C)} + \frac{1}{3} \frac{q_0^2\lambda^2}{D^{*2}(C)} + \cdots\right] \tag{51}$$

in so doing:

$$0 \leqq \frac{dC}{dx} \leqq \infty; \quad 0 \leqq q_0 \leqq q_s; \quad 0 < \frac{q_0}{D(C)} < 1. \tag{52}$$

As the integrated expression contains not a ln but 1/ln we accomplish the formal operation

$$\frac{1}{1 + \frac{1}{2} \frac{q_0\lambda}{D^*(C)} + \frac{1}{3} \frac{q_0^2\lambda^2}{D^{*2}(C)} + \cdots} = b_0 + b_1 \frac{q_0\lambda}{D(C)} + b_2 \frac{q_0^2\lambda^2}{D\lambda^2(C)} + \cdots \tag{53}$$

where coefficients $b_0, b_1, b_2, \ldots$ are determined by multiplying the series in the left and in the right parts of (53). Coefficients b_i in the right part of (53) diminish faster than in expanding (51) and series (53) converges uniformly in [0, 1] interval. Because of the separation of the variables, (48) and (53) give:

$$\left[\frac{D^*(C)}{q_0} + \lambda\left(b_1 + b_2\frac{q_0\lambda}{D^*(C)} + b_3\frac{q_0^2\lambda^2}{D^{*2}(C)} + \cdots\right)\right] dC = dx. \quad (54)$$

Integrating (54) term by term we receive an expression for the determination of C. Note that assuming $\lambda = 0$ in (54) gives the linear variant (50). Substituting $D(C)$ by concrete forms of dependence in (54) and confining ourselves to a certain number of terms we receive various approximations for the determination of flow q through the membrane and concentration profile in it.

Flows and concentration profiles are determined for the diffusion coefficients simulated by dependences (21) and (22). The following relationships have been obtained (the first approximation):

a) $D(C) = D_0(1 + \beta C)$.

The flow is expressed by:

$$q_0 = q_0^* \frac{2l}{\lambda(C_l - C_0) + 2l} \quad (55)$$

where the linear flow q_0^* is equal to

$$q_0^* = D_0\left[1 + \frac{\beta(C_l + C_0)}{2}\right]\frac{(C_l - C_0)}{1}. \quad (56)$$

The concentration profiles are defined by the following relationship:

$$C(x) = \frac{-b + \sqrt{\left[b + 2\beta\left[C_0 + (C_1 - C_0)\frac{x}{l}\right]\right]^2 + 4\beta^2(C_1 - C_0)^2\frac{2(l-x)x}{l^2}}}{2\beta} \quad (57)$$

where C_0 and C_l — concentration boundary values, l — membrane thickness and

$$b = \left[2 - \frac{q_0\lambda}{D_0}\right]. \quad (58)$$

b) $D(C) = D_0 e^{\beta C}$.

An expression is obtained for flow q_0:

$$q_0 = q_0^* \frac{2l}{\lambda(C_l - C_0) + 2l} \quad (59)$$

in which the linear flow q_0^* is:

$$q_0^* = \frac{D_0}{l\beta}\left[e^{\beta C_l} - e^{\beta C_0}\right]. \quad (60)$$

The concentration distribution along the membrane thickness in this case is defined by a transcendental equation:

$$\frac{e^{\beta C_l} - e^{\beta C}}{e^{\beta C_l} - e^{\beta C_0}} = \frac{\lambda(C_l - C) + 2(l - x)}{\lambda(C_l - C_0) + 2l}. \tag{61}$$

Flow expressions $q = f\left(D_0, \lambda\beta, C_0\ C_l, \frac{C_0 - C_l}{1}\right)$ show that when the membrane thickness l decreases and as a result the mean transmembrane gradient $\frac{C_l - C_0}{l}$ increases the flow is limited by a certain limiting value.

Fig. 2 shows concentration profile curves in the points a) and b). In the simplest case as well as when the reaction of the transferred mass is missing

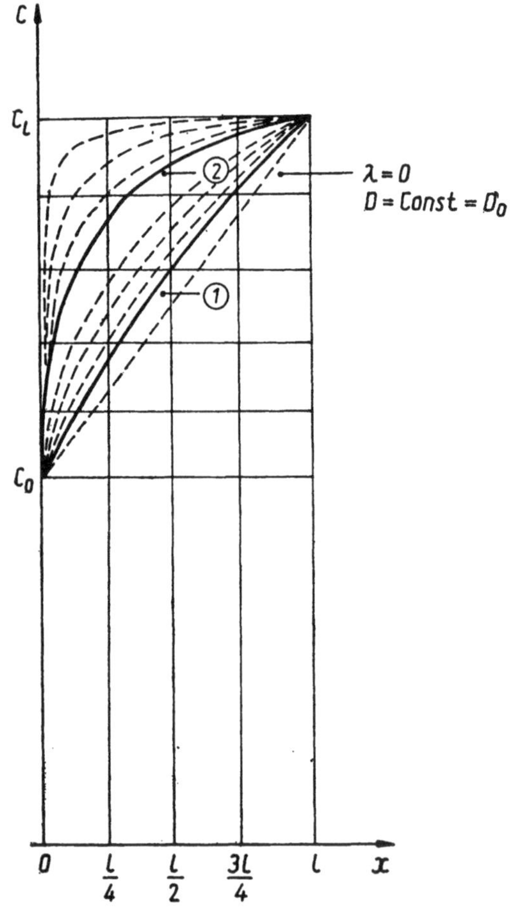

Fig. 2: The distribution of concentrations across the membrane's section in dependence on the form of the diffusion coefficient and values of λ and β parameters.

—— linear model
1 – $D = D_0(1 + \beta C)$
2 – $D = D_0 e^{\beta C}$ $\{\beta = 1\}$
– – – nonlinear models

and the diffusion coefficient is constant:

$$\frac{dC}{dx}\frac{1}{\vartheta}(1-e^{-\vartheta}) = \text{const.} = \frac{q_0}{D_0} \tag{62}$$

or

$$1 - e^{-\lambda(dC/dx)} = \frac{q_0}{q_s}; \quad q_s = \frac{D_0}{\lambda} \tag{63}$$

for positive gradients.

Taking logarithm and integrating (68) give:

$$C(x) = -\frac{1}{\lambda}\ln\left(\frac{q_s - q_0}{q_s}\right)x + C_0. \tag{64}$$

If C_l — is a concentration value on the boundary of $x = l$, then flow q_0 is defined in the following way:

$$q_0 = q_s\left(1 - e^{-\lambda\frac{C_l - C_0}{l}}\right). \tag{65}$$

In this case the concentration distribution in the membrane is linear and coincides with the solution of the diffusion steady equation when the flow corresponds to FICK's 1st law.

As A_0 Fig. 2 we are to note that the concentration curves of small λ parameters corresponding to materials with high permeability are close to the corresponding "linear" curves and coincide with them at $\lambda = 0$. As λ increases, the gradient should be smallest.

The increase of parameter β leads to the distortion of the concentration profiles, and parameter decreasing approximates them to the straight line, corresponding to the linear dependence (64).

6. References

[1] SCHLÖGE, R., „*Stofftransport durch Membranen*", Steinkopff-Verlag, Darmstadt 1964.
[2] KEDEM, O., KATCHALSKY, M., Biochim. biophys. acta 27 229 (1958).
[3] MIKULECKY, D., Biophys. J. 7 527 (1967)
[4] LAKSCHMINARAYANAICH, N., "*Transport Phenomena in Membranes*", Academic Press, New York 1969.
[5] PARLON, R. B., EYRING, H., POLISSAR, M. J., "*The Kinetic Basis of Molecular Biology*", Wiley, New York 1954.
[6] "*Diffusion in and through Solids*" (ed. R. M. BARRER), Cambridge 1941.
[7] DYTNERSKY, U. I., "*Membrane Process of Partition of Liquid Mixtures*", Chemistry, Moscow 1975.
[8] "*Diffusion in Polymers*" (ed. J. CRANK and G. S. PARK), Academic Press, London — New York 1968.

[9] LONG, R. B., HENLEY, E. I., Ind. Eng. Chem., 1965, v. 57, 3.
[10] MARKIN, V. S., TCHIZMADGEV, U. A., "*Induced Ionic Transport*", Science, Moscow 1974.
[11] MALONE, G. H., e. a. J. Theor. Biol. 1972, 36, N 2.
[12] SUCHDEC, S. R., SCHULTZ, I., "*Advance in Bioengineering*", Chem. Eng. Progr. Symp. Ser., №. 114, 67, 1971.
[13] RUBINSTEIN, I. L., Cytology 5, 1973.
[14] RUBINSTEIN, I. L., SELITZER, I. S., Reports of Ac. of Sc. of Latv. SSR. 2, 1973.
[15] LYKOV, A. V., Heat and mass exchange, Energy, Moscow 1972.
[16] BROWN, C. K. et. al., Chem. Eng. Progr. Symp. Ser. 114, 67, 1971.

Bei Fragen zur Produktsicherheit wenden Sie sich bitte an:
If you have any questions regarding product safety,
please contact:

Walter de Gruyter GmbH
Genthiner Straße 13
10785 Berlin
productsafety@degruyterbrill.com